elements of
GRAPH DESIGN

elements of
GRAPH DESIGN

Stephen M. Kosslyn

W. H. FREEMAN AND COMPANY
NEW YORK

Cover Illustration by Paul Mirocha

Library of Congress Cataloging-in-Publication Data

Kosslyn, Stephen Michael, 1948-
    Elements of graph design / Stephen M. Kosslyn.
        p.   cm.
    Includes bibliographical references and index.
    ISBN 0-7167-2263-1. -- ISBN 0-7167-2362-X (soft)
    1. Graphic methods.   I. Title.
QA90.K63 1993
511' .5--dc20   93-27841
                                                  CIP

PRINTED IN THE UNITED STATES OF AMERICA

1 2 3 4 5 6 7 8 9 0 HAM 9 9 8 7 6 5 4 3

contents

preface

This is a how-to book: How to present information effectively in graphs. Right behind the how is the why: The same principles that govern how we make sense of the world also govern how we make sense of graphs, and the advice I offer here is rooted in research results. In this book I show that an understanding of the workings of the eye and mind can be used profitably to design good displays.

This book is not just for researchers and designers. It is intended for department managers making reports who are expected to make use of those fancy (or not so fancy) graphics programs in the company computers; for people in any field making presentations; for students. The reader, I assume, is intelligent and interested—but not necessarily possessed of professional training as an academic, a scientist, or a designer. Nor are the applications limited to today's (or tomorrow's) computer programs; your tools may equally well be a drawing board or pencil and paper. This is graph-making for the people.

This book is a step-by-step guide to constructing comprehensible graphs. We begin by considering the type of data you have and the general point you want to make and select the appropriate type of graph. Then we turn to the practical work of producing the display, systematically constructing the overall framework, the content, and the labels, adding color and texture, and reviewing the outcome. The result is a graph that virtually anyone can read immediately, not a puzzle to be pondered.

The reader interested in the details of the principles of perception and cognition that underlie my graph recommendations will find extensive notes at the end of the book. Moreover, I have provided three appendices, one that reviews basic concepts in statistics, one that summarizes the psychological principles that motivate my recommendations, and another that provides a checklist of features to consider when evaluating a computer graphics program.

I have many people to thank. Steve Pinker played an invaluable role in the development of my approach to graph design, providing insights, constructive criticism, and encouragement. Dr. Susan Chipman—who wrote a request for research when she was at the National Institute for Education, which spurred my interest in graphs —stayed in touch with me over the years and kept my interest in the topic alive; I doubt that I would have written this book if not for her.

Susan also took the time to read and provide invaluable comments on an earlier draft of this manuscript. Jonathan Cobb, senior editor at W. H. Freeman and Company, was superb. He was fully engaged in the project from the outset, gave several versions of the manuscript a careful line-by-line reading, and drew out many of the more interesting ideas that are presented in these pages. Megan Higgins, art director at Freeman, helped me refine my presentation far beyond my original expectations through her ideas about how to organize the material and how to make the book's graphics work effectively. Nancy Brooks's overview and detailed editing were so insightful that I ended up reorganizing the book almost from scratch. Nancy read and thought about every word and stared at every graph, invariably finding the diamond in the rough and seeing how to cut it and polish it.

I also wish to thank Patrick Cavanagh, Lynn Cooper, Richard Mayer, and Armand Schwab for unusually helpful comments on an earlier version of the manuscript, and Charles Stromeyer for pointers into the technical details of research on visual perception. Christopher Chabris, Gregg DiGirolamo, Anne Riffle, Lisa Shin, and Amy Sussman provided all manner of technical assistance. I am particularly grateful to Chris for tracking down so many references and interesting examples of good and bad graphics. And I am grateful to Anne and Lisa for their unflaggingly cheerful management of the myriad details of this project. Chris, Lisa, Billy Thompson, and Adam Anderson helped find the data that are presented in the graphs I use as illustrations, and I thank them for their efforts. And finally, I wish to thank my wife, Robin Rosenberg, and children, Justin, David, and Nathaniel, for their patience and good humor. Justin and David have discovered the joys of graphics on the Macintosh (Nathaniel is still too young for such pleasures), and with luck will find this book useful in the near future.

Stephen M. Kosslyn
August 1993
Cambridge, Massachusetts

how to use this book

For the "big picture" of the many ways in which principles of perception and cognition lead to specific recommendations about graphs (and other displays), read the book straight through, beginning to end. To use it as a workbook or manual to guide you in the design and construction of a particular display, follow the signposts provided in the text.

In either case, begin with Chapter 1, where we see how principles of perception and memory are a two-edged sword. If they are ignored, a display can be uninterpretable; if respected, it can be read at a glance. Move on to Chapter 2, which will help you make the best choice of display format for a particular purpose. If you are using *Elements of Graph Design* as a workbook, be guided by the paragraph at the end of this chapter (and similar succeeding ones) called "The Next Step." Depending on the format you have chosen, "The Next Step" at the end of Chapter 2 will direct you either to Chapter 3 (to make an L- or T-shaped framework) or to Chapter 4 (if your choice is, say, a pie graph or visual table). Now you are beginning the work of producing the graph. If your graph has an L- or T-shaped framework, "The Next Step" at the end of Chapter 3 will send you to Chapter 5 or to Chapter 6 to continue the process. Most recommendations are illustrated with **don't** and **do** examples; if you are willing to trust me, you can simply read the recommendations and look at the pictures, and be guided accordingly.

Having constructed your graph in outline, you are ready to consider further refinements. Chapter 7 provides recommendations for using color, hatching, shading, and three-dimensional effects, as well as for constructing inner grid lines, background elements, keys, and captions. This chapter also provides recommendations for creating multipanel displays.

Chapter 8 shows how people can lie with graphs—warnings for us all. After the display has been created, the recommendations in Chapter 8 can be used to check its overall appearance to ensure that it accurately conveys the patterns in the data. Finally, Chapter 9 discusses how the principles can be used to create effective charts, diagrams, and maps, and offers a brief look at the possibilities for better information displays that are being provided by new technologies.

An important note: The recommendations offered in this book are guidelines, not hard and fast rules. In some cases, aesthetic concerns, context, or your intended emphasis may override my advice. Moreover, the recommendations occasionally may interact in unexpected ways. It is sometimes helpful to generate a rough sketch of several alternative displays that respect the recommendations and evaluate each one. This is particularly easy to do on a personal computer with a good graphics program. The heart of my advice lies in the psychological principles; if you respect them, you cannot go too far afield.

visual table of contents

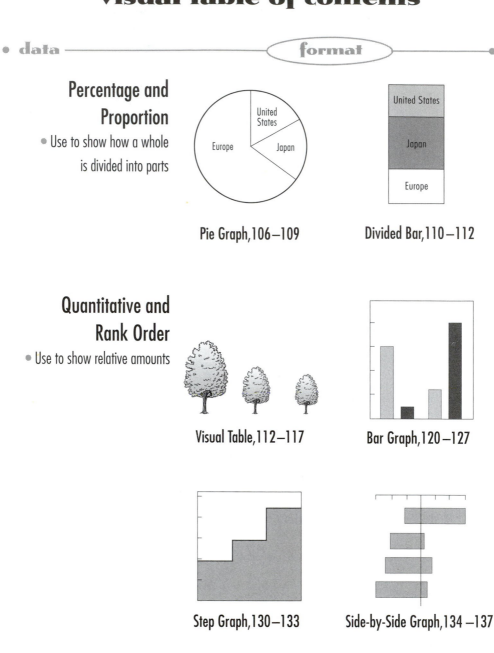
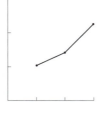
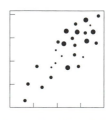

Cumulative Totals

• Use to show how different parts add up to a total

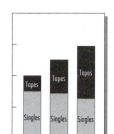

Stacked-Bar Graph, 128 – 129

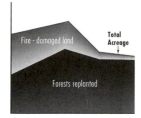

Layer Graph, 150 – 153

Beyond the Graph

• Use to show relationships of value, time, or space

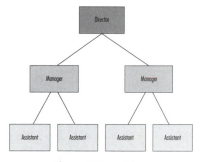

Chart, 238 – 243

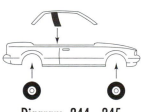

Diagram, 244 – 245

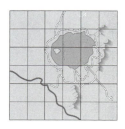

Map, 246 – 253

elements of
GRAPH DESIGN

[chapter]

Looking with the Eye and Mind

1

A picture can be worth a thousand words—but only if you can decipher it. Pictures may be hard to fathom when they are too small or blurry; they are also meaningless, or nearly so, when their content is organized in a way we cannot comprehend easily. The worst offenders may be graphs, which are pictures intended to convey information about numbers and relationships among numbers.

Graphs have become a pervasive part of our environment; they appear in magazines and newspapers, on television and on cereal boxes. Given their near ubiquity, it is surprising that so few graphs communicate effectively. One reason for the abundance of bad graphs is the proliferation of low-cost microcomputers and "business graphics" packages, which often seduce the user into producing flashy but muddled displays. (When was the last time you saw an advertisement for a personal or business computer whose screen did not feature an impressive display?) But although the ease of creating charts and graphs is a major selling point for personal computers, one rarely hears anything about the utility of the displays the machines produce. (When was the last time you could figure out what the display was supposed to mean?) Computers, of course, do not create the problem; they merely multiply it. Confusion and lack of clarity are apparent also in many hand-drawn and traditionally produced graphs.

The obvious question is, why? Why this muddiness in so many instances of such a common form of presentation of information? The answer is the thesis of this book: Many graphs are designed without consideration of principles of human visual perception and cognition. Even most how-to guides for creating charts and graphs ignore the obvious fact that the intended audience are human beings who by their nature have specific perceptual and cognitive strengths and limitations. There is a great wealth of important research about the way we perceive and reason, but it has seldom been mined to aid in the design of good visual displays.

A graph is a visual display that illustrates one or more relationships among numbers. It is a shorthand means of presenting information that would take many more words and numbers to describe. A graph is successful if the pattern, trend, or comparison it presents can be immediately apprehended. Our visual systems allow us to read proportional relations off simple graphs as easily as we see differences in the heights of people, colors of apples, or tilts of pictures mishung on the wall. We are visual creatures, and are good at noting relative differences in sizes, lengths, orientations, and other visually perceptible properties. Graphs can also allow us to appreciate the quantitative relations among multiple elements, and thus can provide us with precise information.

The goal of this book is to help your graphs do their job, which is to communicate effectively. Making a display attractive is the task of the designer, whose talent and visual sense give the graph snap and visual appeal. But these properties should not obscure the message of the graph, and that's where this book comes in. The recommendations offered here will enable you, step by step, to construct the essential elements of an effective display, but the designer's creativity will have plenty of leeway. The recommendations are grounded in facts about the workings of our eyes and minds that affect the way we take in and process visual information; some of these facts reflect properties of the brain that underlie mental functioning. A graph designed according to these recommendations plays to the perceptual and cognitive strengths of those who see it, and avoids being defeated—that is, misunderstood—by the inherent weaknesses of our perceptual and cognitive systems.

Graphics That Work: Three Maxims

Many facts that have been known for decades about how our eyes and minds organize the world around us have direct implications for designing displays; indeed, most of the findings I draw upon are now considered classics. The advice offered in this book is derived from three overarching psychological insights, which can be summarized by three maxims: The mind is not a camera; the mind judges a book by its cover; the spirit is willing, but the mind is weak. Each of these three maxims is an umbrella for a whole class of principles that will come into play, singly or in combination, as we create different kinds of graphs in the next chapters.

The Mind Is Not a Camera

It is natural to assume that our eyes are simple receiving systems that, like a TV camera, register the world as it is. A moment's reflection should convince you that this analogy is misleading: *The mind is not a camera.* Consider the following observations: We readily notice a gain or loss of five pounds on a thin person but may have to strain to see it on someone obese; the box of one brand of soap may look almost twice as big as a competitor's but contains nowhere near that much more soap; a row of reflectors on a dim highway or a formation of geese flying overhead is seen as a single pattern, not as individual objects; former gymnasts watching Olympic gymnastics see much more than do people who are watching the sport for the first time. These phenomena are consequences of the way our eyes and brain work. We are not simply passive receptors, we actively organize and make sense of the world, and when we do so we are at the mercy of the wiring of our eyes and brains.

Although many of the principles we unconsciously rely on are obvious once they are pointed out, some are not. For example, consider the illustration on the following page. If you wear glasses, try taking them off and looking at it; if you do not wear glasses, try looking at it from about ten feet away. You will find it easier to identify the subject as Abraham Lincoln when the picture is out of focus. Why?

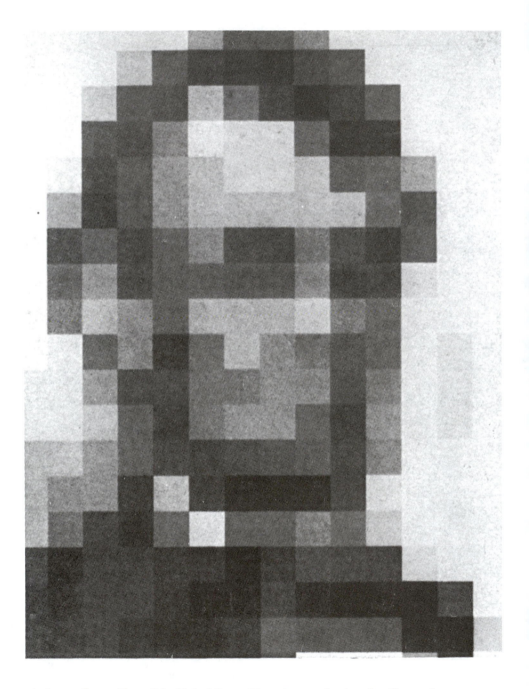

A picture of a president of the United States. To see it more clearly, take off your glasses or step back about ten feet from the page.

When you look around, it seems that you see everything at roughly one level of sharpness, just as would a camera with one lens. But in fact our visual system operates as if it has a number of distinct "input channels." These operate like a set of different lenses, each of which is adjusted to register a different level of detail. Some channels are sensitive only to relatively large changes in a pattern, whereas others are tuned to fine details. The Lincoln figure was created from an image like that on a television screen, which is actually composed of tiny dots of varying degrees of lightness or darkness. The lightness values of the dots were then averaged, making each square a uniform shade. Although the averaging process masks fine variations in the pattern (such as the hairs of Lincoln's beard, his nostrils, and so on), the edges of the imposed squares introduce sharp changes where none previously appeared. Defocusing the picture will not remove any of the actual details of the picture; these were already removed by the averaging process. But defocusing will remove the details caused by the edges of the squares. The useful information from the coarse visual input channels (the overall outline) was being obstructed by information from the fine-detail channels (the edges of the squares); blurring the picture prevents these detail channels from providing spurious input—thus unmasking the information transmitted by the coarser channels. As a result, you see the image better when the picture is blurred.

Why does our visual system operate at numerous levels of acuity at once? Why doesn't it just use the single "best" level? The answer is that there isn't one. Which channel is optimal depends on the task at hand. For counting hairs, the best level of acuity would be the highest; for counting cows, the best would be a lower one, which allows us to ignore extraneous detail.

What do levels of acuity have to do with visual display design? The crucial fact is that a viewer cannot help but pay attention to adjacent patterns if they are processed by the same channel. If you want readers to distinguish at first glance patterns on bars, wedges of a pie, areas on a map, and so on, the patterns should be so constructed that they are processed by different channels; if patterns are processed by the same channel, readers will have to work hard to distinguish them. Researchers have measured the variations in light and dark that lead patterns to be registered by different channels, and their conclusions—discussed in Chapter 3—can be helpful guides to making effective use of cross-hatching, patterns of dashes in lines, and so on.

Here is another example. The brain senses depth in part by exploiting the slightly different images that are registered by each eye. By analogy to a similar process in hearing, this is called stereo vision. The brain reconciles the slight differences in the time a sound arrives at each ear in order to localize the source of the sound; in vision, it reconciles slight left/right disparities in the images in each eye to infer the distance of the object. Stereo vision works well, but it can be tricked to produce an interesting illusion. The illusion occurs because light waves have different wavelengths, which we see as different colors. You can separate out the different colors in ordinary light by holding a prism in front of a window and observing the rainbow that is projected on the wall. The lens of the eye acts like a prism because the eye is aimed slightly to one side when we look straight ahead, and so the lens is angled. Light of different wavelengths is projected to slightly different locations on the retina of the eye just as it is projected to slightly different locations on the wall when we hold up a prism. The result is a "false stereo" effect: We see long-wavelength, warmer colors (such as red) as "closer" to us than short-wavelength, cooler colors (such as blue). This information leads me to recommend that when two lines cross (as in a line graph), the warmer one should occlude the cooler one. If it does not, the back line will seem to be fighting to come forward, trying to snake around the one in front.

Another way in which the mind, unlike a camera, interprets visual elements is by grouping them. There are a number of principles of perceptual organization that describe how we make units of the objects before us.

Proximity. Marks near each other will tend to be grouped together in our minds. For example, xxx xxx is seen as two groups, whereas the same number of elements presented as xx xx xx is seen as three. Thus, depending on how bars in a bar graph are spaced, they are more or less likely to be seen as grouped. Bars in the same group are readily compared. As we shall see, this principle is especially important for associating labels with scales, bars, wedges, lines, and so on.

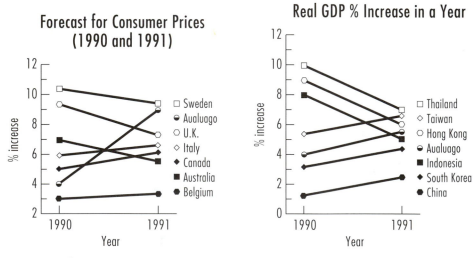

A number of lines can be registered easily if they are parallel, forming a single perceptual group. The difficulty of reading the display increases as more perceptual groups—not more lines—are present.

The three parallelograms embedded in the Star of David.

Good continuation. Marks that suggest a line, even a dashed or dotted one, will tend to be grouped together. For example, — — — — — — is seen as a single unit, not six separate ones; — — — __ __ __, on the other hand, is seen as two units. Bars that are arranged in order of increasing or decreasing size will be more easily apprehended than bars arranged in a way that violates good continuation.

Similarity. Similar marks will tend to be grouped together. For example, — | | | | is seen as two groups. This principle can often be used to help readers to pair corresponding bars, lines, or regions when more than one comparison is being made in a single graph.

Common fate. Lines or marks that seem to be headed in the same direction will be grouped together. Compare the two line graphs on the facing page. A display containing parallel lines is far easier to understand than one containing nonparallel lines; the parallel lines will be grouped into a single unit, whereas the same number of nonparallel lines would be seen as individual units. This principle will play a critical role when we consider how to present complex sets of data in multiple panels in Chapter 7.

Good form. Regular enclosed shapes are seen as single units: [ ] is seen as one unit, but [- is not. Line graphs sometimes produce patterns that are organized perceptually in accordance with this principle.

These grouping effects can be very powerful. Take a second look at a familiar pattern, the Star of David. In this case good continuation, proximity, and good form lead the brain to organize the lines into the two overlapping triangles that we readily see. But there are other geometric figures embedded in the design of the star, which are not obvious without conscious inspection. To find the three parallelograms in the star we must consciously look for them, tracing individual lines to find their perimeters. The parallelograms are no less present in the design than are the triangles, but they are much more difficult to see because no grouping effects, in which eye and mind join forces, are in play to help us see them at a glance.

The Mind Judges a Book by Its Cover

To many Americans, an English accent seems to add about ten IQ points to a speaker and makes what he or she says more persuasive. A second set of principles suggests that just as we can be persuaded by intellectual or verbal appearance, so we are influenced by visual appearance: *The mind judges a book by its cover.* Our visual system and memory system tend to make a direct connection between the properties of a pattern and the properties of the entities symbolized by that pattern. A continuous rise and fall of a line will naturally be taken to reflect a continuous variation in the entity being measured. If the changes in that entity are in fact not continuous but discrete, the continuity implied by a line graph is misleading; a bar graph would better represent the actual situation being depicted. The specific principle here is *compatibility:* The properties of the visual pattern itself should reflect the properties of what is symbolized.

A famous demonstration of what happens when this principle is violated was given by John Ridley Stroop in 1935. Stroop showed people the names of colors written in different colors of ink: For example, the word "red" was written in red, blue, or green ink; the word "blue" in red, blue, or green ink, and so on. When subjects were asked to report the color of the ink, they took more time and made more errors when the word named another color than when it named the color of the ink. Similar interference occurs when people are asked to read the words "large" and "small" written in small and large typeface, respectively. We are impaired when the two messages, that from the physical stimulus itself and that from the meaning, conflict. The brain attempts to fit all inputs into a single coherent pattern and balks when there is conflict. A graph is no place for the Stroop phenomenon.

Perhaps the most fundamental consequence of this class of principles is the observation that "more" of something in a display should correspond to "more" of a substance—higher bars or lines, larger wedges or regions, or the greater extent of a surface should indicate more of the measured material. I once saw an egregious example of a violation of this principle. The artist wanted to convey the burglary rates of different cities and had constructed a variant of a horizontal bar graph; each bar was made up of houses, each house representing a fixed number of houses in that city. In each bar only one house was shown being hit by a burglar—which meant that the *longer* the bar, the *lower* the rate. At first glance, one saw the longer bars as meaning "more" and had to fight the *more is more* principle to realize that they actually meant "less."

The Spirit Is Willing, But the Mind Is Weak

The brain is part of the body, and like any other organ it has its limitations: We can keep only a certain amount of information in mind at any one time. A graph should not require the reader to hold more than four perceptual groups in mind at once. Observe the pairs of displays on the facing page and notice how much harder it is to tell whether the ones on the left are the same as the ones on the right as you progress down the column of pairs. Also notice how you start to try to organize the lines into groups, using the grouping principles just described, after the fourth or fifth row in the figure.

In each pair, is the pattern the same or different? Notice how the decision becomes more difficult as you move down the page.

One of the reasons graphs are useful is that they help to circumvent limitations of the human mind and brain. Although we can consider only about four groups at a time, we can absorb much more information if the information is translated into visual patterns. The table below provides data on blood levels of the (imaginary) fat "parafabuloid" in men and women at two age and income ranges. Is there one group that shows an unusual trend—a tendency for something to increase or decrease—over age? Now look at the graph titled "'Parafabuloid' Level for Age Group." Here it is apparent that all groups but one show lower levels of "parafabuloid" with increased age—women in the lower income group have the reverse trend. Spotting this trend in the table is difficult, but seeing it in this graph is easy: Differences in the orientations of the lines convey the different trends, and the eye and mind quickly register such differences.

But not any graph will do. In the graph headed "'Parafabuloid' Level for Sex," the same data are displayed but arranged differently. In this graph the important conclusion—that there is one group that shows a trend different from the others—is buried. The distinctions are not shown by the slopes of the lines but by differences in the relative differences of the heights of the symbols, which are not easily taken in and compared. The reader is required to analyze the labels and figure out the graph; the information the picture conveys is not helpful. As you lay out a graph, you must decide which is the most important part of the data; it is that variable that should appear on the horizontal axis, since then the differences in the effects of this variable will correspond to differences in the slopes of the lines or variations in the progression of bars produced.

It is a psychological, not a moral, fact that people do not like to expend effort and often will not bother to do so, particularly if they are not sure in advance that the effort will be rewarded. If you expect readers to extrapolate a trend from a bar graph, you are making them do extra work by

"Parafabuloid" Level by Income, Sex, and Age

	Males		Females	
Income Group	Under 65	65 or over	Under 65	65 or over
$0–24,999	250	200	375	550
$25,000+	430	300	700	500

having them connect the tops of the bars in their mind's eye to produce a line they can follow. Why not give them a line graph in the first place? On the other hand, if you want to point out specific values, don't use a line that your readers will have to break up mentally into points; give them a bar graph. In both these cases, the selection of a graph type unsuitable for the data means more work for the reader, and a falling off of your audience.

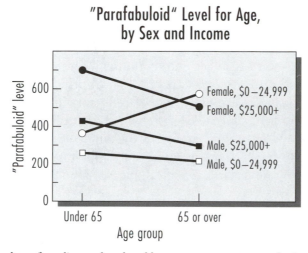

The contrasting slope of one line makes the odd group easy to spot; no such visual cue can be given in a table.

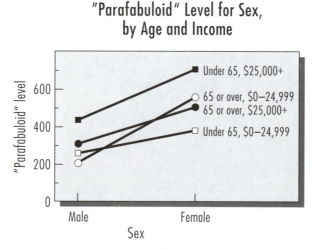

This arrangement of the data makes it hard to detect the different trend shown by one group.

Look at "Nutritional Information per Serving," a horrendous example of overtaxing the reader's perceptual and cognitive systems (and patience). At one point federal regulators considered requiring such displays on food packaging. Try to read this display. Some people never are able to understand it, even if they keep at it; most give up after a first glance.

The problems are many. To begin with, the reader is likely to interpret the small circles in the left panel as pie graphs. Pie graphs divide a single whole into its proportions, but each of these circles does not correspond to a single whole. Rather, the total of four circles in each row is supposed to depict a whole. This arrangement is particularly unfortunate because the circles are positioned closer to the ones above them than to the ones to the sides, and so the grouping principle of proximity leads the viewer to see them as organized into columns when in fact they should be seen as organized into rows. The psychological principles that affect how we see formations of geese and rows of reflectors also operate when we look at visual displays. Our visual systems automatically group things that are near each other into units, and we are likely to see the circles as a line, not as individual shapes. The grouping into columns is misleading and so to read the graph one has to override the initial visual impression, expending time and effort.

The center panel is supposed to convey two different sorts of information: the overall proportion of "Nutritional cont." (actually, vitamins and minerals; the black wedge) and the number of calories (the white wedge)

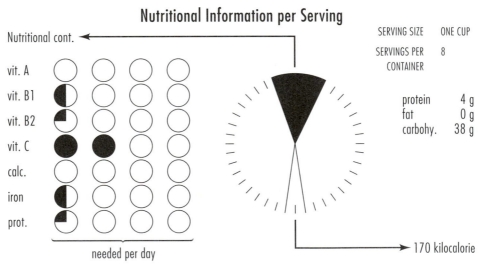

Nutritional Information per Serving

A complex display intended to indicate the nutritive content of food. Try to decipher it.

per serving. (The display uses the technically correct but less familiar term "kilocalorie" to mean "calorie," further confusing the reader.) It is unfortunate that the wedges line up to form a single shape; the grouping principle of good form will lead the reader to think that they are related. It is also unfortunate that the panel looks like the familiar pie graph, possibly misleading the reader to think that the two wedges are different proportions of the same whole. The absence of labels on the arrows forces the reader to work to figure out that they correspond to different things, "composed of" and "produces" for the left and right arrows, respectively. Furthermore, to determine the actual amounts, the reader must count 40 tick marks around the circle (mentally supplying the missing ones); a tedious and time-consuming exercise.

The confusion continues. Notice that protein ("prot.") appears in the leftmost portion of the display and also in the table at the far right; how are these entries related? What is the relation between "170 kilocalorie" (*sic*) and the table at the right about servings? And the scattered organization of the table itself makes it hard to read.

The mind is not a camera; the mind judges a book by its cover; the spirit is willing, but the mind is weak. Each of these maxims subsumes a number of governing principles of perception and cognition. We have already met some, and we will encounter others as we consider what kinds of graphs are best for what kinds of information, and how to design them for maximum effectiveness. To offer specific recommendations, I will draw most heavily on findings about the workings of our perceptual and cognitive systems; to some extent, I will also rely on the results of studies of how people read charts and graphs. I rely primarily on psychological principles of the sort just described not only because the literature on the perception and interpretation of graphs is relatively restricted, but also because many of the studies used flawed displays as their stimuli. It is difficult to interpret a finding that bar graphs are better than line graphs, say, if the graphs were drawn poorly. In addition, the studies vary widely in the tasks they used (asking subjects to compare two portions of a display, compare the parts to the whole, extract specific values, or classify trends), and in the measures that were taken (response times, ratings of sharpness of increases, or accuracy). Thus it is not surprising that the results from different studies sometimes seem contradictory. I have read these studies with a critical eye, and present what I believe to be their essential message (for the interested reader, the literature is reviewed in the endnotes). Finally, I rely primarily on general properties of the eye and mind because I want the principles and recommendations stated here to be general, to help you to design all types of displays—and give you scope to invent new ones.

The Anatomy of a Graph

Graphs not only let us see that there is more of something in one case than in another, they can also tell us how much more. To understand how a graph can provide precise information, we must look a little more closely at its structure.

All graphs, no matter how different individual examples may look, are created from the same components. Typically, they have three primary elements: the framework, the content, and the labels. Thus stripped down, the "graph" in the figure below is reminiscent of the Cartesian graph we met in high school, with its calibrated vertical Y axis and horizontal X axis and origin at 0. In algebra class the content of such a graph was often the curve produced by plotting an equation—a special case of a relationship among numbers.

The framework of the graph sets the stage, indicating what kinds of measurements are being used and what things are being measured. The simplest framework has an L shape, one leg standing for the amount of a measured substance and the other for the things being measured. The vertical leg, the Y axis, usually stands for the measurements (dollars, barrels of oil, degrees of temperature), and the horizontal leg, the X axis, for the things being measured (countries, years, professions).

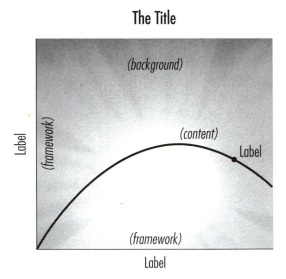

The primary elements of a graph—the framework, content, and labels—with the content shown against a background. You are reading a caption, which explains key terms or draws the reader's attention to specific aspects of the graph.

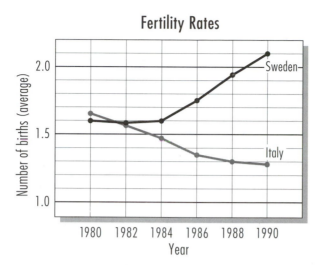

Fertility Rates

This graph has two independent variables, year and country; "Year" has six levels, and "Country" has two.

The content is the lines, bars, point symbols, or other marks that specify particular relations among the things represented by the framework. The positions of content elements are plotted as values along the Y axis (for instance, dollars) and are paired with values along the X axis (for instance, seasons). Graphs display information that is associated with different levels of one or more variables. The line graph above illustrates data—fertility rates—about two variables, year and country. The six levels of the year variable are specified along the X axis, and the two levels of the country variable are displayed as separate content lines. A variable that is broken into several content elements is called a parameter; here the parameter is country, with Sweden and Italy being represented by separate lines.

Each leg of the framework bears a label naming a dependent variable (the type of measurement being made) or an independent variable (the entity to which the measurement applies). Here the dependent variable is the birth rate ("Number of Births [Average]"), calibrated along the Y axis, and the independent variable along the X axis is "Year." Other labels indicate values along the measurement scale (here, specific number of births—1.0, 1.5, 2.0), and the particular entities that were measured (the years 1980, 1982, 1984 . . .). The title of the graph is itself a kind of label. If the content has a parameter, the variable (in this example, country) and its levels also typically are labeled. If the names of the levels—here, Sweden and Italy—are sufficiently distinct, the name of the parameter itself ("Country") can be omitted, as it is in this case.

Graphs also may include a number of optional components. For example, some graphs may include inner grid lines, as does "Fertility Rates." These lines stretch across the framework at regular intervals, either horizontally, vertically, or in both directions. These lines carry no meaning in and of themselves; they simply make it easier to trace along from a point on one leg of the framework to a content element and from there to the corresponding point on the other leg of the framework. Grid lines sometimes can be useful to the reader because, as we shall see shortly, information about location and shape are not automatically combined in the brain.

Some graphs also include a background. For example, a graph about infant development might be superimposed over a picture of a baby. The background serves no essential role in communicating the particular information in the display; if it were eliminated, the display would still communicate the relevant information. Backgrounds can sometimes make a display flashier and more eye-catching, but we will be concerned about backgrounds not so much because of what they can do *for* a display, but because of what they can do *to* a display—make it difficult to understand.

Graphs sometimes include a caption. A caption is a comment on the display, a short description that explains key terms or directs the reader's attention to specific features of the display. Captions are common when displays appear in textbooks, but are seldom seen in magazines or trade books. In some cases, titles and captions are combined, creating a long, discursive title.

This breakdown of a graph into components will help us in two ways. First, from a practical point of view, it allows us to avoid redundancy. Recognizing that a framework, labels, and title are common to most graph types—bar, line, layer, scatter plot—allows the recommendations for producing those elements to be grouped, as they are in Chapter 3.

Second, and perhaps more important, this way of looking at graphs helps us see how graphs work well. To obtain a precise value in a line graph, the reader not only must register the components of the graph but must also relate a specific point on a content element (a line) to the corresponding location on the Y axis. To be effective, therefore, a graph not only must include the essential elements, but these components must also have the proper relations. A graph is more than the sum of its parts; the components must be organized in a way that facilitates our seeing the relations among them. Much of the material in the following chapters addresses this concern.

And this brings me to a cautionary note. There are a very large number of ways in which the components of a graph can be juxtaposed, and there is no way to anticipate the effects—positive or negative—of every possible visual combination chosen to present a particular data set. This observation points up the importance of the psychological principles we shall consider in this book. These principles are universal—they will always apply to all types of displays. Thus even if no specific recommendation applies to a decision you must make, you should keep the principles in mind during all phases of selecting and designing a display. In particular, keep in mind at all times that a graph is intended to make a specific point, and its visual appearance should convey that point. Many of the psychological principles underscore the importance of context, and in some circumstances may contradict a specific recommendation. An example: If your point is to show that a market has become badly fragmented, including many tiny slices in the same pie graph would be a *good* idea—even if each slice cannot be easily discriminated or interpreted. The recommendations are not hard and fast rules but handy guidelines. They usually will be appropriate, but always keep in mind the point you are trying to make and let the clarity of your message be the ultimate arbiter.

The Next Step

Variations in the different components produce a wide range of graph types—line, bar, pie, scatter plot, and more. The next step is to pick one. Chapter 2 presents guidelines to help you decide which type of graph to use in a specific situation. My recommendations for choosing the appropriate type of graph rest on two factors: the particular kind of data you have to present, and the particular types of questions you want the reader to be able to answer. In the course of considering the options, I will be led to discuss additional psychological principles that will be used more generally later in the book.

[chapter]

Choosing a
Graph Format

2

The first step in making a good graph is deciding what kind of graph to use. If the wrong type for the task at hand is chosen, the graph will not communicate effectively, no matter how well it is designed. In order to move toward a choice, you should consider several fundamental questions, not the least of which is whether you should use a visual display at all.

To Graph or Not to Graph?

The following recommendations will help you to decide whether you should use a graph or would be better off with a table or discursive presentation.

● recommendation

Use a graph to illustrate relative amounts.

Use a graph only if the point is to illustrate relations among measurements. As we have seen, graphs use variations along a visual dimension, such as the height of a bar or line over a specific point of the framework, to convey quantitative information. Our perceptual systems allow us quickly to detect differences among heights or the slopes of lines. However, they do not allow us to register absolute heights or slopes very well, and thus graphs are not well suited for conveying specific absolute values. In order to derive a specific value, readers must relate the content to the dependent measure scale, and we are not very good at registering the precise spatial relations of objects that are not physically connected.

Information about shape and information about location are registered separately by two different brain systems and must be combined; error is inherent in the process, resulting in this perceptual limitation. Spatial properties, such as location and size, are processed in the top rear portions of the brain, whereas object properties, such as shape, color, and texture, are processed in parts of the brain that lie under the temples. It is only relatively late in the processing that information about spatial properties and object properties come together in the brain. To register their spatial relation precisely, one must pay attention carefully to two parts of a scene.

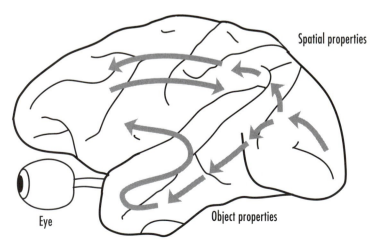

A side view of a primate brain, showing the object-properties (lower arrows) and spatial-properties (upper arrows) pathways.

It often is difficult to pay attention to a content element and the scale at the same time and still see both clearly.

We rarely are aware of the precise spatial relations between parts of a display. I will refer to this limitation—another way in which the mind is not a camera—as the principle of *imprecision*. The strong suit of graphs is the illustration of quantitative relations, and they are not appropriate if you want to convey only precise values: If this is your goal, use a table. If you want to convey both relations among data and absolute values, consider putting a few numbers in critical places on the graph.

● recommendation ───

Specify the subject.

What do you want your readers to know after examining the display? What information will they need?

One useful way to decide what to put in a display is to formulate a precise title. "Plant Productivity, 1940–1990" would lead you to include information different from that in "Number of Units Produced in the United States and Japan." In the first case, you would supply data for each of the different years; in the second, you would simply average the annual production figures to show totals for each country. A display titled "Change in Productivity in the United States and Japan, 1940–1990" would include data by both year and country.

● recommendation ───

Present the data needed to answer specific questions.

A graph is a device to help people answer questions. You must decide what questions readers should be able to answer from the display, and organize the data accordingly.

The linguist Paul Grice formulated rules that ensure smooth discourse between speakers. One of his principles is that people expect a question to be answered with the appropriate amount of information, no more and no less. This principle, which I call the principle of *relevance*, can be extended to visual displays. Readers use displays to answer questions, and they expect to be told as much information as is necessary in the context in which the graph appears—but no more. Relevance is important because, as we have seen, the spirit is willing, but the mind is weak; a robot with an infinite memory and lightning-fast processing speed might not have our limitations.

You should not include any more or less information than is needed to make your point. The principle of relevance is violated when, as often happens, the information in a graph is broken into categories that are extraneous to the message. For example, if you want the readers of your annual report to know whether sales are increasing at the same rates in different parts of the country, plotting the data separately for each of your product lines would only be an obstacle; readers would mentally have to average over lots of lines or bars to obtain the information they needed. Similarly, if variation over years is what is important, average over individual months and do not present the individual months in the graph. On the other hand, it is important to provide enough detail to convey the message. If you want the reader to know about differences in sales of commercial and personal equipment, it would be a fatal error not to include separate bars or lines for each type—there is no way of mentally breaking down a single point or bar into its constituent parts.

• recommendation ───•

Use concepts and display formats that are familiar to the audience.

The display must be designed to communicate to a particular audience. The concepts used must be familiar to that audience, and the display format should be comprehensible to it. This is the principle of *appropriate knowledge:* The reader can know how to interpret a display only if the necessary background is stored in memory. Plotting first or second derivatives and labeling an axis with those terms excludes people who have not studied calculus. Know your audience, and present your information accordingly. Anyone familiar with graphs will equate the notion of "more" with a line rising to the right on a graph, and will understand even without reading the scale that the measured substance is increasing; but even for this simple conclusion to be reached the reader must have the link between "up" and "more" in memory. A more experienced graph reader seeing a display with one rising line and one falling line forming an X pattern can immediately identify that pattern as describing one kind of interaction, that is, a situation in which the value of one variable depends on the value of another. Perhaps goats weigh more than sheep in the summer, but vice versa in the winter. The graph of this situation is an interaction, because the value of one variable (weight) depends on the value of the other (whether it is winter or summer). If season is graphed on the X axis and weight on the Y axis, the line for goats will cross the line for sheep (provided that the animals weigh about the same amount on average), forming an X pattern. By taking time to analyze the picture presented by the graph, any reader will eventually come to the correct conclusion; but only readers who have seen and analyzed a number of such graphs will be able to associate the X pattern itself with a meaning stored in memory and thereafter know the significance of the pattern immediately.

To understand something, we must grasp its implications and notice its relations to other things. Visual displays often are more like stories than like pictures of objects; although they have components that must be identified, it is the relations among the components that convey the specific information. This aspect of understanding in part requires recalling the rules of how a display format works, and in part requires reasoning.

After you have decided what numbers to present, and have thought about your audience, you are ready to consider particular formats. In this chapter we will discuss the most familiar types of displays, which are standard in most computer graphics programs. (There are many other types of displays that are less well known. In the future, computer graphics packages almost certainly will expand one's options, providing greater opportunity both to communicate clearly and to confuse the point.) Your choice of graph format will depend on the kind of information being conveyed and the kinds of questions you want your reader to be able to answer.

Graphs for Percentage and Proportion Data

How are operating costs divided among salaries, energy costs, benefits, and so on? What is the contribution to total profits of each division? What percentage of the national Republican vote came from each region of the country? These are questions about how a whole is divided into parts; numbers of this sort must add up to a fixed total, 100% for percentages, 1.0 for proportions.

Percentage and proportion data can be presented using a variety of formats, including bar graphs and line graphs. However, if you want to emphasize that the components sum to a single whole, then you should choose a pie or divided-bar format. These formats make clear not only the relations among the various components but also their relation to the whole. (If you want to display components of a whole that do not add up to a constant amount, see the discussion on stacked-bar graphs and layer graphs below.)

Pie graphs are circular, the proportions of the whole indicated by the sizes of wedges; divided-bar graphs are rectangular, the proportions of the whole indicated by the sizes of internal segments. Pie graphs do not have a labeled scale, but rather represent amount by variations in the size of content elements; divided-bar graphs can have a labeled scale. In both formats, the framework is implicit in the external boundary of the content. For these kinds of displays, the reader can easily extract precise amounts only if they are specified by separate labels.

When to Use a Pie Graph

The most common way to display how a single whole is divided into parts is to use a pie graph. The relative area of each wedge represents the proportion of the whole taken by the component.

•recommendation ───

Use a pie to convey approximate relative amounts.

A pie graph effectively conveys general information about proportions of a whole. However, if the reader is supposed to obtain relatively precise amounts, such as the percentage of one part, it is better not to use a pie; this information cannot be easily obtained because it is difficult visually for us to measure the angles, chords, or areas of wedges precisely. It is very difficult, for example, to use the **don't** graph to see that there were 13% more car and mobile home loans than bank and finance loans: The principle of *perceptual distortion* is at work. Some dimensions are systematically distorted by our visual systems; specifically, our visual impression of area is less than what is actually present (the mind is not a camera). The underestimation is more severe for larger regions. This is a problem with pie graphs because about one-fourth of graph readers apparently focus on relative areas of wedges when they read such graphs. Of course, one way to convey precise amounts is simply to label the wedges, putting the numbers in them (if space permits) or next to them. But this converts the pie into a table, and loses the power of a visual display.

•recommendation ───

Use an exploded pie to emphasize a small proportion of parts.

If the reader is supposed only to make approximate visual comparisons, the pie format provides a particularly good way to draw attention to a small percentage of the total number of components. An exploded pie display is constructed by displacing the important slice or slices, as if a wedge of pizza had been pulled out from the pie.

Nerve cells in the visual system can be thought of as "difference detectors"; they respond most strongly to a change in stimulation. We focus first on features of a display that are brighter, darker, in motion, or are otherwise different from the surrounding parts. This is the principle of *salience*, which states that large visual changes will be noticed first (again, the mind is not a camera). An exploded slice or two that disrupts the outer contour of the framework will be noticed immediately.

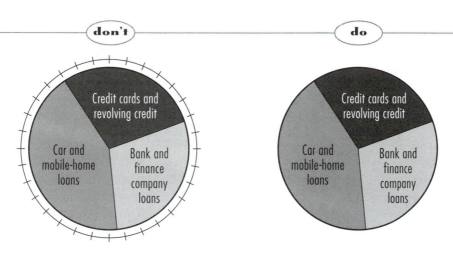

don't

do

Do not use a scale with pie graphs; the reader will have to struggle to count the number of ticks. In this graph, credit cards and revolving credit are the topic of interest, and hence this information is most salient.

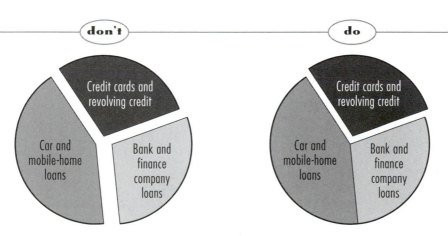

don't

do

Use an exploded pie only when a distinct contour can be disrupted by exploding the wedges; the two intact pieces of the pie on the right define a contour, which makes the exploded wedge stand out.

If proportions vary greatly, do not use multiple pies to compare corresponding parts.

Compare the two pies in the **don't** version. How did Peugeot and Renault do in the European Community (EC) in general (left) compared with Italy in particular (right)? This is difficult to fathom, in part because the relevant wedges are in different locations in the two pies—a shift that cannot be avoided when the proportions vary widely in the two pies. Now compare the two **do** pie graphs; it is clear that there is a rough correspondence between the revenues per region (left) and employees per region (right). It is easy to compare multiple pies when the wedges are in roughly the same positions in each, but when the corresponding parts are in different locations, the reader is forced to search for them one at a time. Remember, the spirit is willing, but the mind is weak. Many readers may simply give up.

Therefore, if the proportions are very different, use the pie format only if the point is to show that there is a major change over the levels of the parameter. This recommendation is rooted in the principle of *informative changes*, which leads readers to expect any change in a pattern to mean something. The brain registers differences and tries to interpret them. When things stay the same, there is no new information; when something changes, there is—or should be—new information. Readers expect every noticeable change in the appearance of parts of a display to mean something; if it does not, the change is simply a distraction. We are not passive viewers, but rather build up expectations about the world as we see and actively seek specific information—which requires effort.

In short, if the proportions are very different across levels, *and* the reader is supposed to compare specific components, a bar graph is the preferable format.

Car Markets

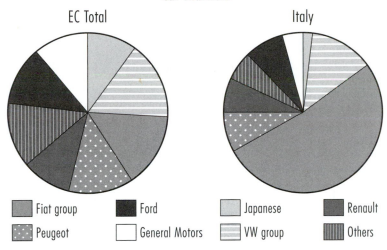

Revenues and Employees per Region of Electrotechnical Corporation

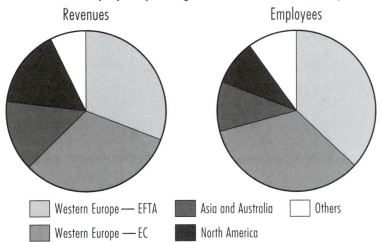

Corresponding wedges are hard to compare if, as in "Car Markets," they are not in corresponding positions.

When to Use a Divided-Bar Graph

The divided-bar graph is a kind of square pie, the length of each segment within a bar representing the proportion of one component of the whole.

• recommendation ——————————————————————————————————————
Use a divided bar to convey accurate impressions of parts of a whole.

Unlike area, distance along a single extent is not distorted by the visual system, and the reader can gain an accurate impression of amounts by noticing the height of a segment. Moreover, a reader can extract precise amounts in this format if direct labels are included in the display. But direct labels lose the power of a visual display; they require adding and subtracting rather than comparison of amount or extent. To convey a sense of the amount of each segment, use a scale with divided-bar graphs.

However, a disadvantage of using a scale with divided-bar graphs is that readers must realize that the absolute height of a segment above the X axis usually is misleading; it is the vertical length of an individual *segment* (the distance between its bottom and top) that is important. In the example, the relative value of recreation is not 68%, but 41%. Readers must count and subtract numbers to get exact percentages—they cannot simply read off numbers. If the subtraction is too difficult, the process will violate the principle of *limited processing capacity* (the spirit is willing, but the mind is weak). These limitations may arise from basic properties of our brains. A good graph can be read at first glance; it is not a puzzle to be solved.

Graphs for Quantitative and Rank-Order Data

Considerable data in our personal and business lives answer the question, How much? How many units were produced, how much money was spent, how many people were involved? To answer this kind of question, visual tables, line and bar graphs, side-by-side graphs, step graphs, and scatter plots are the most useful formats, especially when you are not interested in the relation between individual amounts and the sum total. These formats can be used to convey rank-order or interval information (numbers on a continuous scale, such as money, weight, or temperature). They also can be used for percentage and proportional data *if* you want to emphasize comparisons among relative amounts; they are less useful for communicating how the amounts add up to a fixed whole.

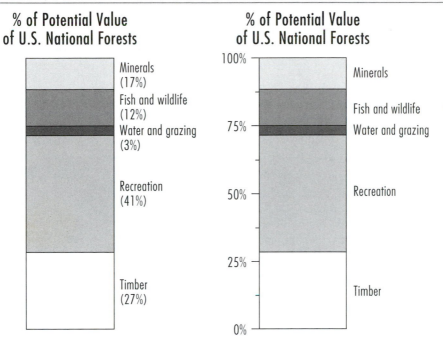

If you want the reader to know precise amounts, label the segments directly. A scale is accept-able if you want to give the reader an impression of the actual amounts of each segment, not to convey precise values.

When to Use a Visual Table

Visual tables are pared-down graphs, which convey information by properties of the content alone; objects are drawn so that their sizes or numbers vary in accordance with the amount being represented. Some visual tables use pictures of actual objects, and some use abstract content such as bars. A special case is the isotype, in which bars are created by repeating small pictures, each of which corresponds to a specific amount of the measured substance. Isotypes are particularly interesting because different pictures can indicate component parts of the measured entity; for example, one could specify the number of male and female employees in the different divisions of a company by showing rows of pictures of a man or woman next to a label for each division. Each picture would stand for an increment of, say, 100 employees, with more pictures (forming a longer row) indicating more of that type of employee in that division.

A visual table does not have a framework and hence has no scales; the information is conveyed solely by the relative sizes of regions.

• **recommendation** ───
Use a visual table to convey impressions of relative amounts.

If the reader is expected to gain only a general sense of the ordering of the levels of the independent variable, then a visual table is appropriate. These displays eliminate unnecessary material, in accordance with the principle of relevance. A set of water bottles, varying in size, may be enough to illustrate the increase in consumption over time (as in the **do** graph). But because visual tables do not have scales, and because the visual system tends to distort area, visual tables cannot be used to depict trends with precision (as is mistakenly attempted in **don't**).

However, you can label each content element with its amount (in the same way that you can label the sizes of individual pie wedges), allowing the elements to convey precise information. For example, if the sizes of balloons (presumably filled with hot air) indicate the number of speeches delivered by each of several politicians, you could put the name of the politician on each balloon and have a tag with the actual number of speeches attached to the string. But be careful: This practice can easily result in a cluttered display, relegating the content elements to the role of mere decorations. If you want to convey precise numbers, consider simply using a tabular format instead.

don't

Bottled Water Sold in Yuppieville

do

Bottled Water Sold in Yuppieville

1980   1985   1990
1 cm = 1,000 gallons

1980   1985   1990

Our visual systems register relative areas, but not very accurately; it is too difficult to use a scale to extract precise amounts from a visual table.

When to Use a Line Graph or a Bar Graph

Line graphs and bar graphs are often used to display changes over time. They are the most common formats used to display quantitative data. The choice between them depends in part on the nature of the entities being measured and in part on the question you want readers to answer with the display.

Bar graphs have the standard L-shaped framework and use bars as content elements. The heights of the bars specify discrete amounts.

Line graphs are just like bar graphs, except that a line is drawn instead of bars. The height of the line at each point indicates the amount, and changes in the height of the line indicate changes in amounts.

• recommendation ——————————————————————————————————————

Use a line graph if the X axis requires an interval scale.

The continuous rise and fall of a line is psychologically compatible with the continuous nature of an interval scale, one which specifies the actual amounts along a continuous measurement scale. The yearly pattern of a young man's fancy is more clearly perceived from **do** than from **don't**. Time, temperature, and amount of money are measured using an interval scale.

• recommendation ——————————————————————————————————————

Use a line graph to display interactions over two levels on the X axis.

One of the reasons it is so difficult to predict real-world events is that multiple variables interact: The effects of one can be understood only in the context of others. For example, the effects of a presidential election on real estate values in Washington, D.C., will depend in part on the type of property. In prosperous sections of the city, values may increase for all but the most expensive homes, whereas in poor sections they may not be affected at all. This interaction, between neighborhood and type of housing, reflects the dependence of the value of one variable on the value of the other. The units of line graphs—lines and patterns—make them ideal for displaying interactions in the data, because the quantitative pattern is signaled directly by the visual patterns themselves. Most experienced graph readers can immediately recognize that a sideways V pattern (with the ends of two lines converging) indicates one sort of interaction, whereas an

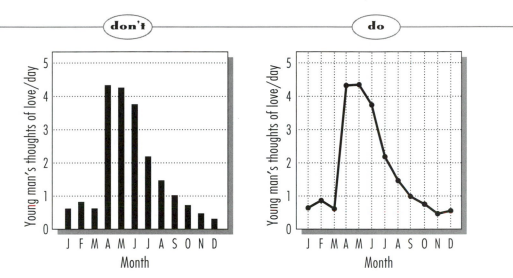

The continuous variation of a line is compatible with the continuous variation of time; if you want the reader to note precise point values, put dots or symbols along the line.

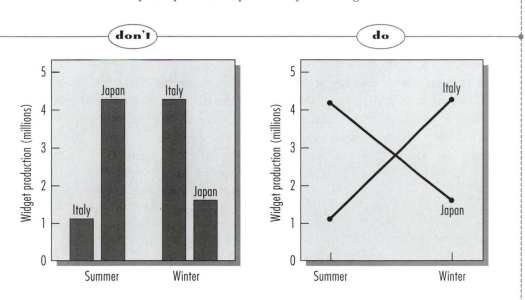

Experienced graph readers can interpret typical patterns of lines at a glance.

X indicates another. As in **do**, we could graph widget production on the Y axis, season (summer versus winter) on the X axis, and country (Italy, Japan) as separate lines. The crossing lines indicate that in Japan more widgets were sold in summer than in winter, but vice versa in Italy; in contrast, if the lines formed a V on its side, that might indicate that there was no difference between the countries in summer, but fewer widgets were sold in Italy and more widgets were sold in Japan in the winter. On the other hand, in order to see bars as unified patterns that convey specific types of interactions, one must mentally connect the ends of the bars, expending time and effort.

If there is little chance that readers will improperly infer an interval scale along the X axis, then a line graph is a good way to convey interactions even if a nominal or ordinal scale is used along the X axis. Nominal scales specify names, not amounts. The numbers on the jerseys of football players are members of a nominal scale. These members often have no necessary ordering (although sometimes an order is imposed by usage), so you are usually free to organize them so that the display is visually simple. Ordinal scales give rank orderings, such as the order in which runners finish a race. Rank-order scales ignore the difference between contiguous values; the difference in time between first and second place may be much larger than between second and third, but this is not indicated in the scale. For further distinctions among interval, nominal, and ordinal scales, see Appendix 1.

● **recommendation**

Use a line graph when convention defines meaningful patterns.

Patterns of lines can signal specific information for readers who have had experience with similar line graphs (and so have appropriate knowledge). For example, results from the Minnesota Multiphasic Personality Inventory, a standardized personality test, are graphed as scores on nine subtests. Although these subtests are members of a nominal scale (they have no inherent order), they appear in a conventional order along the X axis. The higher someone scores on each subtest, the higher the point over that location on the X axis; by connecting the points with lines, the graph assumes a jagged appearance. To an expert, the distinctive patterns of peaks and valleys thus formed are signatures of specific maladies.

Line?

don't

do

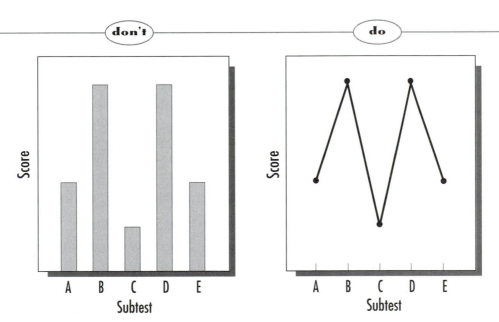

If data are presented in a standardized format, readers familiar with the subject can interpret patterns of lines at a glance.

• recommendation ─────────────────────────────────────

Use a bar graph to show relative point values.

Use a bar graph if the reader is supposed to compare specific measurements. If you want to show how much more one division produced than another for each of three years, you would have three pairs of bars along the X axis—a pair for each year—as shown in **do**. The height of each bar specifies a point value, and this value is relatively easily read. If a line graph is used for the same data, the reader must locate and perceptually isolate a point along a line and note its height along the Y axis; this process requires careful allocation of attention and greater effort. You should not require the reader to reorganize a pattern into a new set of perceptual units (in this case, break a line into a set of points) to answer the necessary questions. This is another example of the principle of limited processing capacity.

• recommendation ─────────────────────────────────────

Use a bar graph if more than two values are on an X axis that does not show a continuous scale.

Look at the **don't** version, in which the slope of the line makes it appear as if the income for consulting partners is accelerating rapidly. This visual impression is less striking in **do**. If the intervals of space along the X axis indicate ranks or levels of a nominal scale, not actual equivalent intervals between years, a line will give a misleading impression. Bars, however, do not suggest continuous intervals, and are appropriate if you want readers to see point values or make these specific comparisons. In general, if you have three or more levels on an ordinal (ranks) or nominal (names) scale along the X axis, and patterns of lines have not assumed meaning via conventional usage, then a bar graph is appropriate. According to the principle of compatibility, the properties of the pattern itself should be consistent with properties of what is symbolized. If a line varies in height continuously, so should the measures of the entities represented along the X axis. The principle of compatibility leads us to use a bar or step graph if the entities do not vary continuously and more than two of them are placed along the X axis.

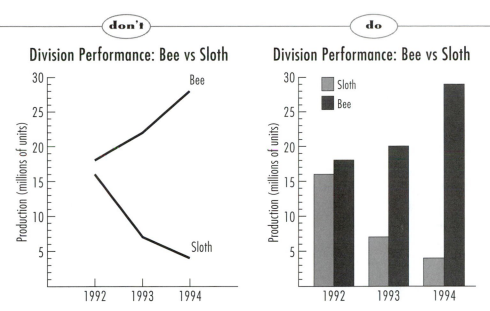

The heights of bars define specific points, whereas lines specify continuous variations. It is more difficult perceptually to break up a line into points than to detect the tops of bars.

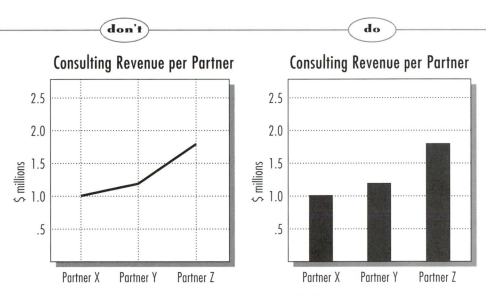

A line graph for these data inappropriately suggests a rapid rise along a continuous variation.

Let reality decide between vertical and horizontal bars.

According to the principle of compatibility, the properties of the pattern itself should not conflict with the properties of what is symbolized. If a graph stands for things that actually have a left-to-right or top-to-bottom order in the world, the graph should symbolize them with bars that have the same spatial arrangement, as in **do**.

A mark at the top should always stand for top, and one at the bottom for bottom; one at the left for left, and one at the right for right. Imagine how hard it would be to drive if instead of a steering wheel we had a lever that had to be pushed forward to get the car to turn right and pulled back to get it to turn left.

Use a horizontal bar graph if the labels are too long to fit under a vertical display.

If no other recommendations are violated, use a horizontal bar graph if the labels are long. If a vertical format requires novel abbreviations, the principle of relevance may be violated: Readers may receive less information than they need in order to decipher the material effortlessly.

When in doubt, use a vertical bar-graph format.

Vertical formats are more familiar to most people than are horizontal ones, and increased height may be a better indicator of increased amount; all cultures recognize "higher" as "more," but some cultures, particularly in the Middle East, do not recognize extension from left to right as "more." The common associations of a culture should be respected when designing a display—this is the principle of *cultural convention*. This principle is an offspring of the maxim that the mind judges a book by its cover; unlike the principle of compatibility, however, the interpretation of cultural patterns is not inborn but is learned, and so may vary in different societies. In Western culture, increases commonly are indicated from left to right and clockwise around a circle, green or blue means "safe" and red means "dangerous," and left is "liberal" and right "conservative." But these associations are not worldwide, and the conventions of your audience should not be contravened. If learned associations stored in memory conflict with the message of your graph, processing will be impaired.

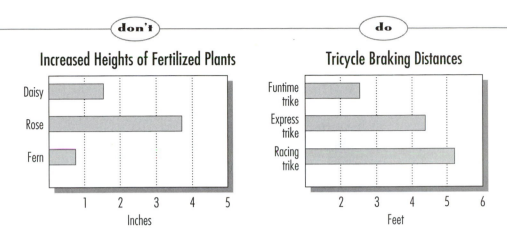

don't

Increased Heights of Fertilized Plants

do

Tricycle Braking Distances

The properties of the pattern itself should not conflict with those of the objects or relations being represented.

When to Use a Side-by-Side Graph

A side-by-side graph shows pairs of values that share a central Y axis; its framework is thus a T or inverted T. To the left are measures of one level of an independent variable (say, one type of tax), and to the right are measures of another level of that variable (another type of tax). Each pair of bars splayed out from the center illustrates a level of a second independent variable (for example, country).

• recommendation ──────────────────────────────

Use a side-by-side graph to show contrasting trends between levels of an independent variable.

Look at "Taxes": In which country is the disparity between the two classifications of taxes greatest? A side-by-side bar graph dramatically displays the relative patterns over different levels of an independent variable. Recall that the visual system readily registers differences, and so readers will quickly note the difference between a regular progression on one side and an irregular one on the other. However, note that the principle of appropriate knowledge may be violated because one set of bars that actually indicates "more" extends greater distances to the left; readers familiar only with Cartesian graphs, in which values to the left of the vertical axis are negative, may be confused. Be sure your audience knows how to read these displays.

• recommendation ──────────────────────────────

Use a side-by-side graph if comparisons between individual pairs of values are most important.

Because the bars in a side-by-side graph are immediately juxtaposed, it is easy to notice differences in their lengths as violations of symmetry. The example illustrates the difference in vitamin consumption for two types of guppies; it is immediately obvious in **do** that fan-tailed guppies need disproportionately less B_{12} than other vitamins, compared with spiny-tailed guppies. Side-by-side bar graphs allow the reader not only to contrast trends but also to flag specific differing values. In these displays it is easy to spot the few right–left pairs that are asymmetrical in a sea of symmetries, or the few that have longer bars to the left when the rest have longer bars to the right.

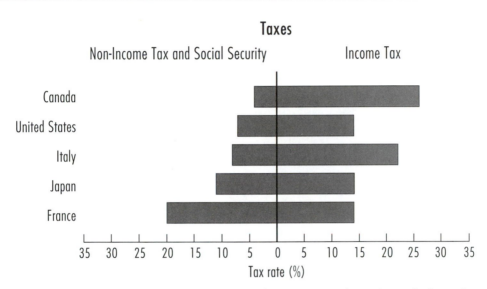

Canada has the largest disparity in the two types of taxes—a point that is clear only if a reader knows how to interpret this format.

don't   do

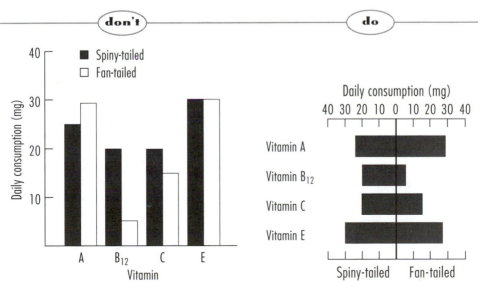

The relatively large asymmetry for B_{12} is easy to spot immediately in a side-by-side format.

If differences between paired values are important, illustrate them directly.

Both versions of the "Lateralization" graph illustrate the output from a computer model of the functions of the left and right sides of the brain as the two sides might develop if one has particular experiences and innate biases in how information is processed. In **do** the white portions of the bars indicate the "strength" of certain processes in the left and right sides, and the black portions of the bars indicate the degree to which one side is "stronger" than the other. It is clear at a glance that black bars can extend to the left or right side, which indicates that different sides of the brain can be better at specific processes (if the model is accurate), depending on one's personal experiences and processing biases. If your goal is to illustrate a difference in paired values, readers should not be asked to subtract the value of the bars mentally. Why make the readers work harder than they have to (and risk losing them)? It is preferable to present the difference (shown here in black) along with the data. (Recall the principles of limited processing capacity and relevance.)

Avoid using side-by-side displays for more than two independent variables.

Consider a side-by-side display of the average caloric intake (the dependent measure) of male and female gulls for each of twelve months (the independent variables). Say that months are marked off along the axis, so that there are 12 pairs of bars, and data from male gulls are represented by bars going to the left, and data from female gulls by bars going to the right. This display—the **do** version—is readable. But if we add a third independent variable, age (young versus old), each bar of the previous display must be replaced by a pair of bars, one for "young" and one for "old" birds. The virtues of side-by-side bars are quickly lost when the arrangement becomes this complicated and we no longer can apprehend at a glance a visual pattern that conveys a message. Unless the data produce regular visual patterns, this format is not well suited for graphing more than two independent variables.

don't do

Lateralization of Individual Subsystems with Inhibition Model

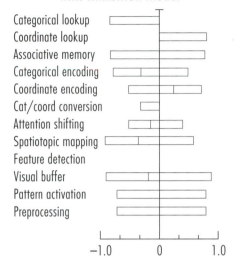

Lateralization of Individual Subsystems with Inhibition Model

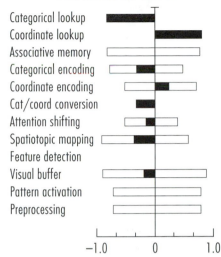

The black portions of the bars allow the reader to see immediately the differences in the total extents to the left and right.

don't do

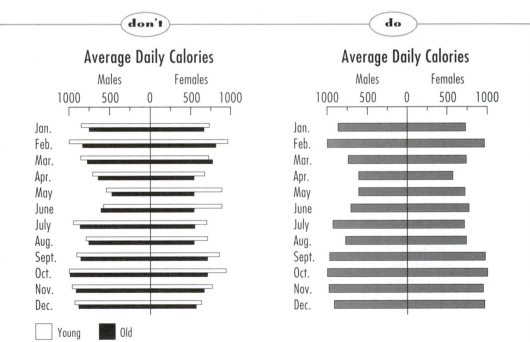

If the pattern is too complex, the reader will not easily detect differences in symmetries.

When to Use a Step Graph

Step graphs, like line graphs, use lines as content elements, but these lines trace out horizontal plateaus that change height in discrete jumps (as do the heights of bars in bar graphs). These graphs are like bar graphs in which the bars are pushed together to form a set of steps and all demarcations but the upper line are eliminated. As in bar graphs, the heights of the lines indicate discrete amounts.

Step graphs illustrate quantities, rank orders, or even percentages and proportions (provided that you want to emphasize comparisons among relative amounts, and not the relation of each component to the whole).

● **recommendation** ───

Do not use a step graph for two or more variables or levels of a single variable.

From the **don't** graph it is almost impossible to tell the percentage of automobile sales for the different countries. Each independent variable or level of an independent variable is represented by a different series of steps. If the overall values of the two or more levels are similar, the steps will cross and will be difficult to distinguish because of perceptual grouping. Alternatively, if one level has much greater values than another, it will not be immediately clear whether the upper series of steps corresponds to the cumulative total or simply to the value of one of the levels. In these situations, present the data in separate panels, as shown in **do**.

● **recommendation** ───

Use a step graph to illustrate trends among more than two members of nominal or ordinal scales.

A step graph is useful if the readers are supposed to notice relative changes over three or more values on the X axis *but* these values do not vary continuously (if they do, use a line graph; if there are only two values, a line graph is acceptable even for nominal and ordinal scales). In many ways, a step graph is a good surrogate for a line graph for entities that are not arranged on an interval scale: Pushing the bars together creates a pattern that may allow the experienced reader to take in a trend at a glance. The only drawback to these displays is that a bit more effort is required to isolate individual point values, and so a bar graph (where spaces between the bars can isolate the point values) is more appropriate if this emphasis is your goal.

don't   do

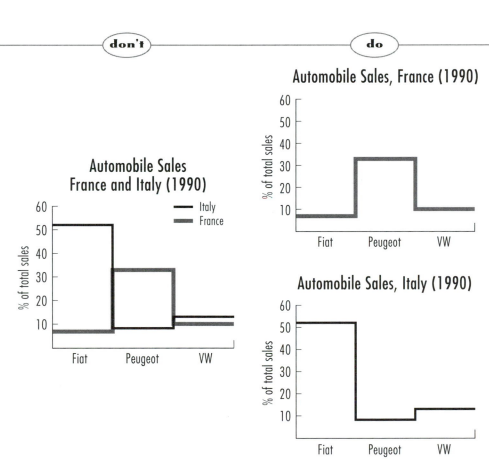

A combined step graph is difficult to read and easily mistaken for a layer graph (which would illustrate a cumulative total).

When to Use a Scatter Plot

Scatter plots, which have the standard L-shaped framework, employ point symbols (such as dots, small triangles, or squares) as content elements. The height of each point symbol indicates an amount. These displays typically include so many points that they form a cloud; information is conveyed by the shape and the density of the cloud.

• **recommendation** ────────────────────────────

Use a scatter plot to convey an overall impression of the relation between two variables.

Scatter plots are appropriate if you want readers to obtain only an overall impression of the relation of two variables. For example, if the heights and weights of 100 people are plotted, you see a cloud drifting up toward the right, indicating that the two variables are positively correlated. You also see that this correlation is not perfect; some very tall people are skinny, and some short ones rotund. Readers who are familiar with these displays can extract a reasonably good sense of how closely values on one variable are related to values on the other. Do not include a grid, as in **don't**; the point is to convey an overall trend, not individual values. Scatter plots often include far more points than can be grouped and seen at once; therefore, they are not a good idea if readers are supposed to discern specific point values (use a table instead). Use a scatter plot in strict accordance with the principle of relevance; it is easy to overwhelm a reader with too much information in these displays.

• **recommendation** ────────────────────────────

Avoid illustrating more than one independent variable in a scatter plot.

It is possible to depict different independent variables in a scatter plot by using different symbols to mark the points for different levels of the parameter. For example, say you want to illustrate the relationship between the price of a car and its weight, for U.S. and foreign vehicles. You can clearly see this relationship for both categories in **do**, because the distribution of symbols falls into separate perceptual groups. But data are not always this cooperative: Where the two levels are winter and summer sales of *all* cars, U.S. and foreign (**don't**), the distribution is such that the different symbols are hard to distinguish and read. It usually is better to plot different independent variables in different scatter plots. The only exceptions to this recommendation occur if you want to show that data from two independent variables are intimately related (and so the fact that the points cannot be distinguished is itself compatible with the message being conveyed), or if the clouds formed by different sets of points can be easily distinguished because they are in different parts of the display.

46

don't   do

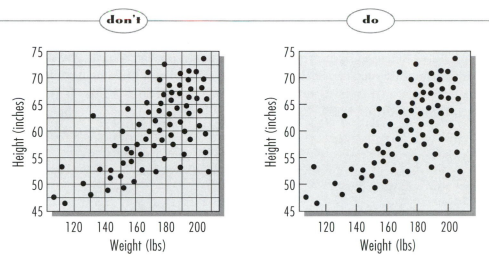

Scatter plots convey an impression of trends in the data; they generally are not well suited for conveying values of individual data points, and an inner grid is usually a distraction.

don't   do

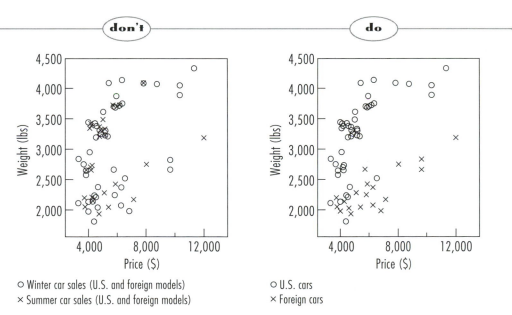

○ Winter car sales (U.S. and foreign models)   ○ U.S. cars
× Summer car sales (U.S. and foreign models)   × Foreign cars

Two levels of a parameter can be easily detected in a scatter plot only if their symbols occupy different parts of the display.

Fit a line through a scatter plot to show how closely two variables are related.

In many cases, scatter plots are used to illustrate a trend, which can be summarized effectively by a line. Unlike the line in a line graph, this "fitted" line does not connect up sets of points. Rather, it falls through the "center"—the most populated area—of the cloud of points. The most popular way to fit such a line requires finding the location that minimizes the average distance of the points to the line (see Appendix 1). When a best-fitting line is provided, the average distance of the points from the line illustrates visually how tightly the points are related by the two variables being graphed. At left is a scatter plot without a best-fitting line; the scatter plot on the right shows the same data with such a line. It is clear that the best-fitting line helps the reader to discern the trends in the data.

Graphs for Cumulative Totals

It is sometimes useful to graph cumulative totals, particularly when a quantity varies over time. Stacked bars and layer graphs are used to convey information about how different kinds of things add up to a total, when the total may vary from case to case. For example, one could use segments of a bar to illustrate the amount of taxes paid by the branches of a company in different countries, and have separate bars for each year; the overall heights of the bars would change, along with the heights of each segment. Stacked bars are similar to divided bars, the content indicating components of a total amount. Unlike pies and divided bars, which always add up to a constant whole (100%), the cumulative total varies; also, stacked bars often have an explicit L-shaped framework. Like ordinary bar graphs, the framework specifies a measurement scale (on the Y axis) and levels of an independent variable (on the X axis).

Layer graphs (also sometimes called layer charts, segmented graphs, or surface graphs) are similar to line graphs, except that the height of each line indicates a cumulative total. They specify how one type of entity and its components vary continuously. The content is a set of layers, each of which indicates the amount of a particular component of the whole. The layers might be the expenditures of different departments in a plant (accounting, production, marketing, and so on). In this case, if the Y axis specified expenditures and the X axis the month, each line would represent the cumulative total for all departments under that line—and each layer (the space between a pair of adjacent lines) would indicate the expenditures from just one department.

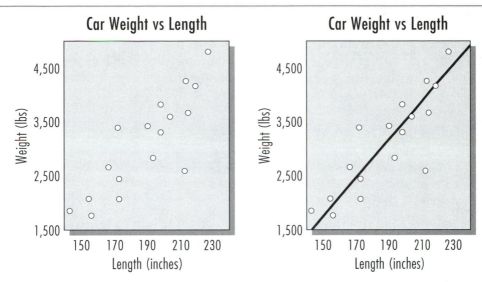

A best-fitting line serves as a summary of a trend and helps readers to see the strength of the correlation between two variables.

When to Use a Stacked-Bar Graph

Stacked bars consist of two or more parts, each of which specifies the value of a different level of the independent variable. Stacked bars should be used only to illustrate components of wholes that change; they are most effective if the following recommendations are respected.

• **recommendation** ─────────────────────────────
Use stacked bars if the wholes are levels on a nominal scale.

Stacked-bar graphs have many of the same properties as divided-bar graphs; indeed, the only real differences are that overall height of the stacked bar varies, and that typically several stacked bars, illustrating levels of two or more independent variables, are used. The **do** graph uses stacked bars to break down the total number of widgets sold in four cities by industrial and home uses; the overall height indicates total number of widgets. Thus, the recommendation pertaining to divided bars also applies here (see pages 28–29). This format is desirable if the independent variable of most interest (different cities, in the example) does not specify a continuous quantity; it is useful to have spaces between the bars for the same reason that bar graphs are preferable to line graphs when there are multilevel nominal scales along the X axis (see pages 36–37).

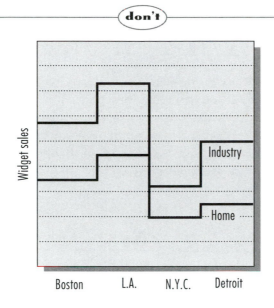

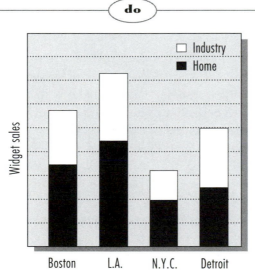

Stacked bars clearly illustrate the contributions of each component. The top arrangement of these data is not easily read and could be mistaken for a step graph.

When to Use a Layer Graph

A layer graph is like a set of stacked bars that are pushed together. These displays illustrate changes in components of a whole over time or another interval scale along the X axis. The recommendation pertaining to divided-bar graphs (pages 28–29) applies here as well as the following ones.

• recommendation ───

Use a layer graph only if the X axis is an interval scale.

If the X axis depicts a continuously changing variable—that is, values on an interval scale—the flux of the lines that define the layers will be compatible with the content, as in **do**. But if the X axis is an ordinal scale (one that specifies ranks), or is a nominal scale (one that names different entities) as in **don't**, the eye will incorrectly interpret the quantitative differences in the slopes of the layers as having meaning (use stacked bars instead). The principle of compatibility suggests that layer graphs be used only if the values on the X axis are arranged along an interval (continuous) scale.

• recommendation ───

Use a layer graph to illustrate changes of parts.

Look at "Trends in Federal Funding": Which research component was changing most rapidly? Layer graphs are useful if you want to illustrate the relative change in one component over changes in another variable. Because the spaces between lines can be filled, they can be seen as shapes (the visual system organizes regions of common color or intensity into a single shape) and the change in a single proportion can easily be seen.

• recommendation ───

Do not use a layer graph to show precise values of parts.

The overall height of the individual lines in a layer graph indicates a cumulative total. To extract a specific amount, readers must note the relative heights of the lines defining a layer at a specific point along the X axis. Given our limited processing capacities, this information is difficult to derive; to make a precise comparison, we must subtract heights mentally while keeping in mind the results of previous subtractions.

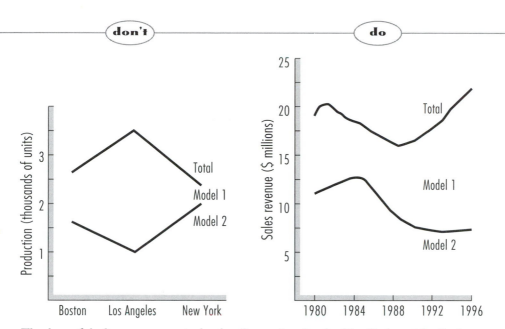

don't do

The slopes of the lines over a nominal scale—Boston, Los Angeles, New York—misleadingly suggest a continually varying quantity.

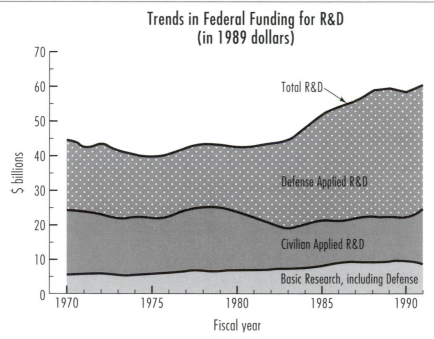

Trends in Federal Funding for R&D
(in 1989 dollars)

This graph is particularly effective because only one component—Defense Applied R&D—changed dramatically over time and therefore it is easy to see the differences in the changing widths of the segments.

Multiple Panels

Any of the graph formats described in this chapter can be used in multipanel displays, which include two or more graphs. There are two types of multipanel displays. Mixed multipanel displays include two or more different formats; the individual formats are selected independently for each set of data, according to the recommendations offered above. Pure multipanel displays include two or more graphs of the same type. The following recommendations pertain to pure multipanel displays; the individual displays in mixed multipanel displays should be selected in accordance with the previous recommendations. (In Chapter 7 we consider how to create both types of displays.) The question becomes, then, when should you divide the data up and display it in separate panels?

• **recommendation**

Do not present more than four perceptual units in one panel.

If there are more than four bars over each point on the X axis, or more than four groups of lines, it is best to divide the data into subsets and graph each subset in a separate panel (in Chapter 7 we will consider how to decide what goes in each panel). If a display is too complicated, we cannot readily understand it because we cannot hold enough of it in mind at the same time, as is the case in **don't**. If a company has six different departments, it is usually better to present separate graphs for, say, U.S. and foreign expenditures than to try to cram all that information into a single display. The one exception would be if many of the lines have similar slopes, so that they are perpetually grouped into relatively few units (see pages 6–7).

When we "hold something in mind" we are in fact retaining information in short-term memory. To understand a display fully, one often must draw deductions and notice implications, which requires reasoning; these reasoning processes operate on information in short-term memory. Short-term memory has a relatively small capacity, placing severe limits on what we can understand. Many poor displays are incomprehensible in part because they overload the viewer's short-term memory capacity.

A perceptual unit, defined as such partly by the principles of perceptual organization, is known as a chunk. Much research indicates that we can hold in mind only about four chunks at the same time; this is the principle of *limited short-term memory capacity* (the spirit is willing, but . . .). Because a chunk is a psychological, not a physical, unit, we cannot determine whether a bar or a line graph should be broken into multiple panels simply by counting marks on the page.

don't

do

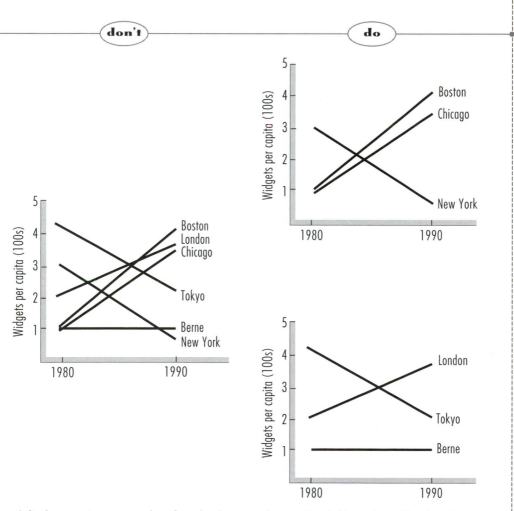

A display containing more than four chunks cannot be apprehended in a glance. Break such displays into separate panels.

Use multiple panels to highlight specific comparisons.

You may sometimes want readers to compare particular subsets of data. Presenting these data in separate panels will cause them to be grouped via proximity and hence lead the reader to compare them. For example, if sales growth is presented for various countries, data for the countries that are to be compared should be plotted together, as in **do**.

Presenting complex data in multiple panels not only decreases the load on short-term memory but also simplifies the depicted patterns. If the display is too complex, the resulting pattern probably will not be familiar—and so will not match a pattern stored in memory and will not convey meaning in a glance. If a complex display is broken into a number of simpler ones, many of the resulting patterns typically can be read without having to note the precise values of individual points. In the following chapters we will discuss how to break data into groups, each of which should be plotted in a separate display.

Optional Features

Finally, let us consider a number of features that can be included in most of the display formats discussed: a key, error bars, inner grids, the use of two dependent measures, a caption.

Use a key when direct labels cannot be used.

In many cases it will not be clear whether a key is necessary until the rest of the display is produced; only then can you determine whether there is space to label each wedge, object, or segment directly, or whether a key would reduce clutter.

To understand a key, the reader must memorize the associations between the names and the content elements, which requires work and may tax our limited-capacity short-term memories. If at all possible, don't use a key. There are two exceptions to this general recommendation, both of which are illustrated in **do**: Use a key when there are so many wedges, objects, or segments in a small space that is impossible to label them directly (the labels would not group properly or would have to be so small as to be unreadable); or when the same entities appear in more than two graphs of a multipanel display (and hence a key will reduce clutter and the load on short-term memory).

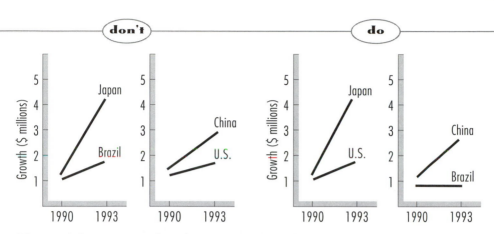

The intended comparisons—here, between two industrialized and two nonindustrialized nations—are facilitated by appropriate grouping.

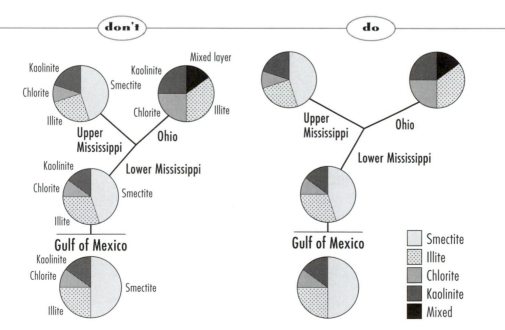

A key can not only reduce clutter but also make a display more visually interesting.

Typically include error bars.

Graphs are often used to illustrate averaged data. An average, by defini-tion, is a measure of the "central tendency" of the numbers. The same average can be obtained from an infinite number of sets of data: 10 is the average of 9 and 11; of 3, 5, 15, and 17; and of −62, −56, 1, 45, 50, and 82. In this case, the variability in the numbers increases for each of the three successive sets (see Appendix 1 for a discussion of averages and common ways to compute variability). It often is useful to convey information about the variability in a data set, because readers should have less confidence in the average if the variability is high. In an opinion poll, we have less confidence in the numbers that have large "margins of error" (a measure of variability).

If a display is intended to report relative point values or a trend, it is usually a good idea to include, as in the examples shown, a small "I" bar (an error bar) centered on each point, the top and bottom of the bar indicating the range of plus-and-minus one standard error of the mean (or some other measure of variability; see Appendix 1).

Like all my recommendations, this one must be taken in context. De-pending on your purpose, error bars may not be appropriate. Error bars were not included in the previous displays, nor will they often appear in later chapters, because of the principle of relevance. Because these graphs are designed to illustrate specific recommendations, not to present data, including error bars would provide more information than is appropriate, and their presence would only detract from the point being made.

Use an inner grid when precise values are important

Use an inner grid if you want the reader to extract specific values, which can be read off the Y axis more easily if one can trace along grid lines. The grid lines serve to connect points along the line or heights of bars to specific places on the axes, taking advantage of the principle of proximity to mitigate the imprecision of our visual systems.

This line graph illustrates a special case in which grid lines are particu-larly helpful. Look at the vertical difference between the lines over 5 and over 2; which difference looks larger? In fact, they are the same distance apart, as you will see if you use the grid lines to compare the two differ-ences. Our visual system tends to see the minimal distance between the lines, not their difference in height. Grid lines help the eye to focus on the vertical extent itself.

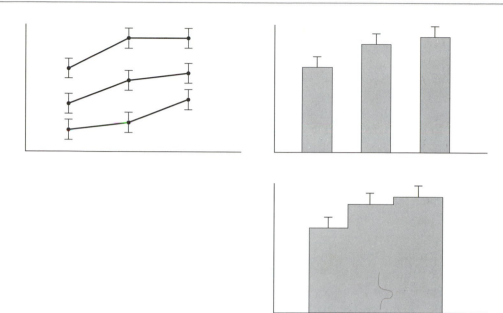

Error bars indicate the range over which measurements are likely to vary with repeated sampling, helping the reader to know which differences to take seriously.

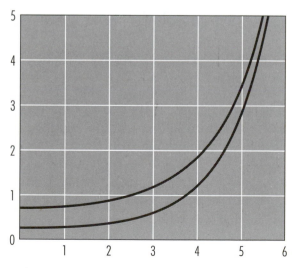

Without an inner grid it is almost impossible to tell that the lines have the same difference in value on the Y axis over the 2 and 5 values of the X axis.

● **recommendation**

Graph two dependent measures in a single display only if they are highly related and must be compared.

As a rule, do not graph different dependent variables in the same display; in most cases, graph the different data in separate panels of a multipanel display. Graphing two dependent variables in the same display forces the reader to keep track of what goes with what, taxing our limited processing capacity. The only exception to this recommendation occurs when the dependent variables are intimately related and their interrelations are critical to the message being conveyed, because plotting the data in the same display allows the principles of perceptual organization to group them into a single pattern. The volume of oil pumped and the failure rate of a piece of pumping equipment are probably connected, and it makes sense to graph these two dependent measures in one display, as in **do**. On the other hand, oil volume and the CEO's frequent-flier mileage (**don't**) are not as immediately related, and the graph is not particularly helpful.

● **recommendation**

Do not use mixed bar/line displays to show interactions.

As is evident in **don't**, it is more difficult to see interactions if a mixed display is used because bars and lines do not group to form simple patterns (they are too dissimilar, and so the principle of similarity will not be effective). For easiest comparison of trends and interactions, use multiple lines.

● **recommendation**

Include a caption to clarify unfamiliar terms and specific features.

A caption is an explanatory note accompanying the display. The principle of appropriate knowledge leads you to ensure that the reader knows the terms necessary to understand the graph, and in some cases labels should be clarified in a caption. If the display is complex, note explicitly in the caption which patterns in the display are of most interest.

The Next Step

The next step is to begin to create the display. If you have selected a pie graph, divided-bar graph, or visual table, go to Chapter 4; if you have selected any of the other types, which have an L-shaped or T-shaped framework, go to Chapter 3.

Options?

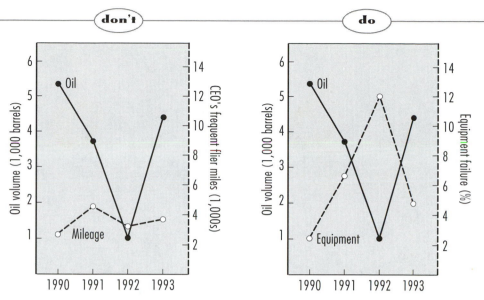

Plotting two related variables together can reveal useful information, such as the impact of equipment failure on the amount of oil that was pumped.

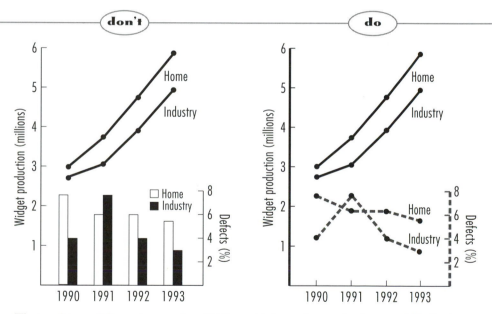

The trend toward decreasing number of defects with increasing production, especially for the industrial widget, is not immediately evident from the mixed line/bar display; the visual system more easily groups different lines into a pattern than it groups lines and bars.

[chapter]

Creating the Framework, Labels, and Title

3

The different graph formats share many features. All have labels and a title; most (except pie graphs, divided-bar graphs, and visual tables, discussed in Chapter 4) have an L- or T-shaped framework. In this chapter, recommendations are presented for creating the framework, labeling it, and titling it. Once the framework is labeled and titled, turn to the appropriate chapter—4, 5, or 6—to create the content elements; to label them, follow the recommendations in this chapter.

Creating the L-Shaped or T-Shaped Framework

The framework of a bar, line, stacked-bar, step, layer, or scatter-plot graph is an L shape. The Y (vertical) axis indicates the amount of what was measured (the dependent variable) and the X (horizontal) axis indicates the things to which the measurements apply (levels on an independent variable); the roles of the vertical and horizontal legs are reversed in horizontal bar graphs. Side-by-side graphs, which are like two horizontal graphs that share the same Y axis, have a T-shaped framework. First we will consider general recommendations about the overall form of the framework; we will then turn to the details of producing the X and Y axes.

● **recommendation** ──

The framework should be easily detectable.

According to the principle of *detectability*, marks must be large enough and drawn with lines that are easily seen, as in **do**. Brain cells (neurons) that detect changes in one region of space inhibit other brain cells that detect changes in nearby regions, and so if a mark is not large or heavy enough, the cells that typically would detect it are inhibited from responding (the mind is not a camera). The limit of detectability depends on a wide range of factors, such as the color of the figure and its background, the contrast of intensity of light and dark elements, and the shape of the figure. It is impossible to provide hard and fast rules to ensure that a mark can be noticed, because so many factors affect each other.

● **recommendation** ──

The parts of the framework should be grouped to form a single unit.

The legs, or axes, should be connected, so that by the principles of perceptual organization the framework will be seen as a single whole, as in **do**. If the legs are not connected, and the gap is too large to be completed perceptually by good continuation, the reader will have to work to figure out the display.

64

don't   do

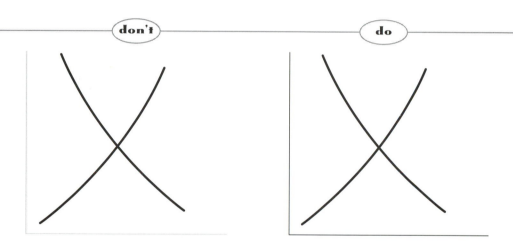

If the framework is difficult to discern, the content functions as a visual table.

don't   do

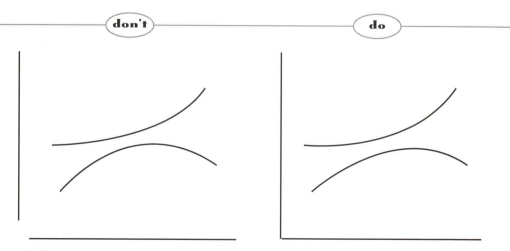

Confusion can ensue if the segments of the framework do not form a single unit.

• **recommendation** ——————————————————————————————————————

The height-to-width ratio of the axes should make differences in content discriminable.

The ratio of the height to the width is the aspect ratio. If the X axis is too long relative to the Y axis, differences in bar heights (or line heights, for line graphs) may not be readily discriminable. All bars drawn in the **don't** framework would be relatively small, and the percentage of difference not large enough to be easily discriminated. Adjust the aspect ratio so that in accordance with the principle of compatibility, actual differences in the data produce corresponding visible differences in the display.

• **recommendation** ——————————————————————————————————————

For accurate visual impressions in a line graph, duplicate the Y axis at the right.

A curious visual illusion causes people to underestimate the right-hand point along a line if it has to be compared against a Y axis on the left-hand side; this misapprehension is amplified if two or more lines are graphed in the same framework. The error is not very large—at most about 8%—and may not be important for many purposes. But if you want the reader to gain an accurate visual impression, draw the Y axis on both sides of the graph, as in **do**.

Note that the use of an inner grid is still a good idea if you want the reader to obtain specific point values; moreover, if you use an inner grid, there is no need for a second Y axis. For aesthetic purposes, you can always "box" an L-shaped framework, as in **do**.

don't   do

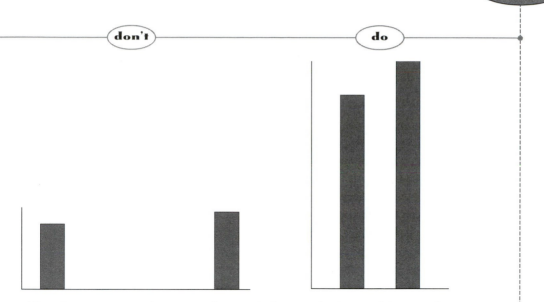

Adjust the aspect ratio so that, in accordance with the principle of compatibility, significant differences in the data produce corresponding visible differences in the display.

don't   do

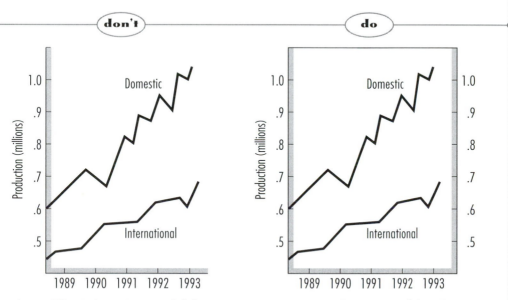

A second Y axis (or an inner grid) helps to convey an accurate visual impression of the relation of two lines at the right side of a graph.

If two dependent measures are plotted, use different colors or patterns for the two Y axes.

Two different dependent measures, each on a separate Y axis, should be presented in the same display only if you are emphasizing a close relation between them. Line graphs are usually the most appropriate type of graph for such information because lines produce the most easily identifiable patterns. In accordance with the organizational principle of similarity, use the same color or pattern to plot the line and the corresponding Y axis, as in **do**, allowing the viewer to see immediately which Y axis goes with which data. When using colors, choose ones that are well separated in the spectrum, and make sure that adjacent colors have different brightnesses. Use warm colors to define a foreground; avoid using red and blue in adjacent regions; and avoid using blue if the display is to be photocopied. (For a further discussion of color, see Chapter 7.)

With two dependent variables, put one Y axis on the left and one on the right.

To promote discrimination and proper grouping with the content, put one Y axis on the left and one on the right, as in **do**, rather than putting both on the same side. Putting both axes in the same area can result in incorrect grouping of labels with axes.

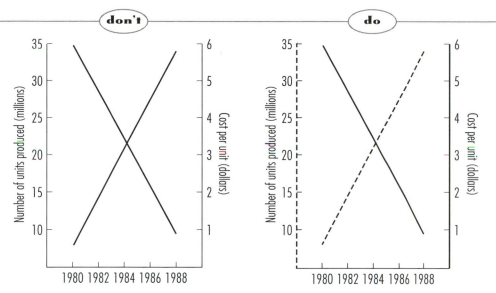

A difference in line patterns allows the reader to see which Y axis goes with which content line.

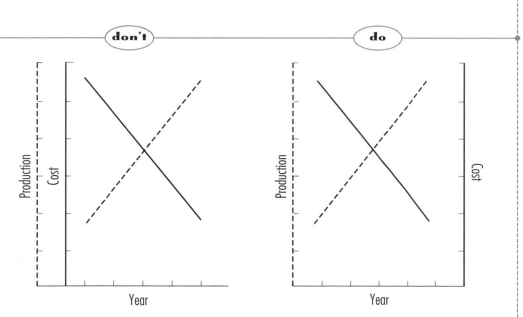

Proximity improperly groups two Y axes on the same side of the graph with each other and the inner axis with both content lines.

Creating the Axes

Four questions need to be addressed in order to produce the appropriate X and Y axes. First, if there is more than one independent variable (for example, age and income level), which should label the X axis, and which should be treated as parameters, represented by separate bars or lines? Second, how should the levels of the variables along the axes be ordered? Third, what range of values should be included along the framework? Finally, how should tick marks be produced?

For convenience, in the following recommendations I will assume that the X axis is used to present an independent variable (company, country, political party, etc.) and the Y axis is used to present the dependent variable (dollars, amount of oil, number of votes, etc.). In horizontal graphs, these roles are reversed; if your format is horizontal, treat recommendations for the X axis as recommendations for the Y axis, and vice versa.

● **recommendation**

Put the most important independent variable on the X-axis, and treat the others as parameters.

Which *team* had more scored points, Giants or Redskins? This should be easier to determine from the graph on the left. But if you want to know how the *points* were made—which team had more scored points than allowed points—the version shown on the right is preferable. The difference between the two is the choice of independent variable to put on the X axis and the corresponding parameters (the different bars). Because our visual systems group things that are close to each other, the two graphs make different sorts of information easiest to detect at a glance. The choice of which independent variable to put on the X axis depends in large part on the purpose of the display: The most important independent variable should be on the X axis. That variable will be perceived as the most important factor, the others as the background information.

To determine which variable is the most important, write down a concise description of what you take to be the main point your graph should summarize. In the football example, perhaps your summary is: "The Giants had roughly equal numbers of scored and allowed points, whereas the Redskins had many more scored points." Here the important contrast is between teams, which is the subject of the sentence and the first element mentioned, and your graph would look like the one on the left. On the

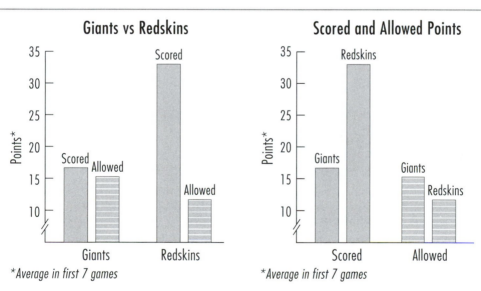

Depending on the variable chosen for the X axis, different patterns emerge and different trends and interactions are easy to see.

other hand, you might write: "More scored points were made by the Redskins than the Giants, whereas more allowed points were made by the Giants than the Redskins." Here the focus is on the difference between scored and allowed points, suggesting that this is the most important variable, and leading you to use the arrangement of the graph on the right.

This recommendation is based on the principles of relevance and limited processing capacity. If an inappropriate organization is used, the reader must mentally transform the graph to extract the information you are attempting to highlight; the reader can easily see differences between bars at adjacent points along the X axis, but must work to locate distant corresponding bars and remember what to compare with what. This effect is even more dramatic with line graphs, where the orientation of the line conveys information about variations of levels of the independent variable on the X axis, and differences in the relative heights of the lines at each point convey information about the variations of levels of the parameters (see pages 10–11).

● **recommendation** ——

When in doubt, put an interval-scaled independent variable on the X axis.

If there is no clear distinction between the importance of the variables, is one of the independent variables on an interval, that is, a continuously varying, scale? If so, put that variable on the X axis, as in **do**. The progressive variation in heights from left to right along the X axis will then be compatible with the variation in the scale itself; such changes in height are easily detected by our visual systems. If there is more than one independent variable with an interval scale, it usually is best to put the one with the greatest number of levels along the X axis. For example, if you are graphing the number of people of different ages who voted Democratic in Somewhere County in different years, put year on the X axis if you have ten years and only two age groups, and vice versa if you have only two years and ten age groups. This procedure cuts down the number of separate content elements and thereby reduces the load on our short-term memory capacities.

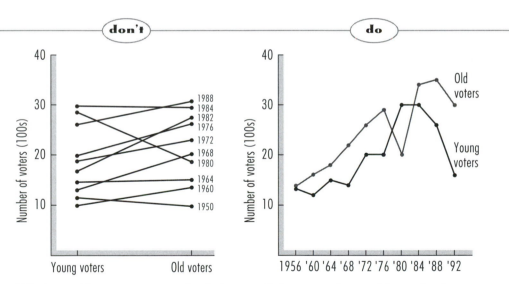

don't   do

If both variables are on an interval scale, it is usually best to put the one with more levels on the X axis. This choice produces fewer lines and, usually, a simpler pattern.

All else being equal, put on the X axis the variable that produces the simplest pattern.

If none of the foregoing recommendations applies to a particular case, put on the X axis the independent variable that allows you to make the simplest pattern of content elements; this arrangement will be the easiest to apprehend, as you can see by comparing **don't** and **do**. This recommendation is easiest to carry out if you use a computer to create the graph and can experiment with different ways of producing the display.

Creating the Scales

Once you have decided which independent variable should be placed along the X axis, the levels of that variable must be ordered. If the levels reflect measurements along an interval or ordinal scale (years, weight, age, placement in a race), this is not an issue; the levels are ordered from smallest to largest (left to right or bottom to top). However, if the levels are names of things (such as countries, players, products, and so on) that are not drawn from an interval or ordinal scale, they can be ordered in many ways. If there is no conventional or common-sense order for the things (if you are graphing the lifetime income of members of the folk group Peter, Paul, and Mary, the order is obvious), consider the following recommendations.

Use position to indicate greater or lesser quantities.

Spatial dimensions have a psychological beginning, symbolized in a graph framework by the intersection of the X and Y axes, by convention the site of the origin. Locations farther from that beginning should signal increased amounts. The principles of compatibility and cultural convention suggest that quantities (such as time or volume) should increase from bottom to top, left to right along a line (as in **do**), or, in the case of a pie graph, clockwise around a circle.

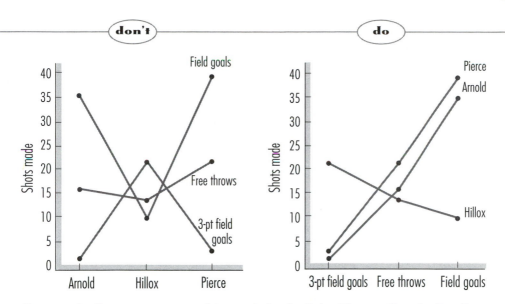

Fewer crossing lines are one measure of the complexity of a display. However, if crossing lines form a familiar pattern, they can be comprehended more easily than unfamiliar uncrossed lines.

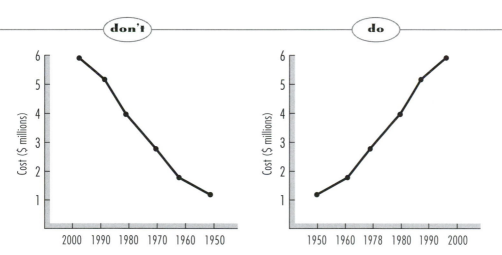

In Western culture, a pattern moving left to right will be automatically, and in this case improperly, interpreted to indicate "more."

• recommendation ————————————————————————————————
Display locations should be consistent with actual locations.

The principle of compatibility suggests that data about actual positions in space should be presented in the corresponding positions in the display, as in **do**. I have (too often) seen this principle violated in graphs showing properties of the left and right sides of the brain, where "left" is on the right side of the X axis and "right" is on the left side. This forces the reader to reorganize the display mentally, a task that requires effort and is sometimes prone to error.

• recommendation ————————————————————————————————
Order a nominal scale so that cases to be compared are adjacent, or so that the simplest pattern is produced.

Both **don't** and **do** illustrate the percentage of 18–19-year-olds who were full-time students during the academic year 1987–88. The **do** version is much easier to take in at a glance; you immediately see the ordering and differences among the countries. If you want the reader to compare specific cases—that is, levels on a nominal scale—put their steps or bars next to each other. If no specific comparisons are particularly important, by the principle of informative changes a simple progression of increasing height from left to right is to be preferred over a jagged series if the ordering along the X axis is arbitrary; otherwise the reader may assume that each jag has a greater significance than it actually does and waste time figuring out what it is (here the only meaning is that some things have more than others, differences measured by the heights of the steps). If there is no particular point being made and the variables on the X axis have no intrinsic order, then order the levels along the X axis to produce the visually simplest pattern. This pattern can be organized effectively and apprehended easily in accordance with the principles of perceptual organization.

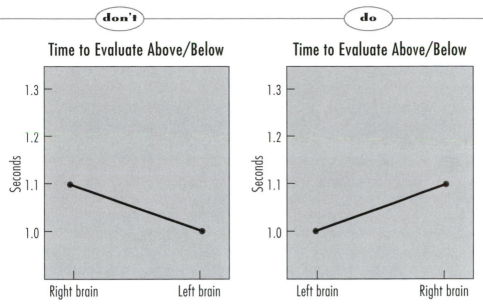

The left side of something should be illustrated at the left end of the X axis, and the right side should be illustrated at the right end of the X axis.

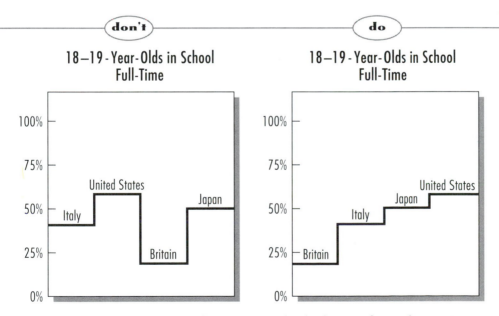

If no particular comparison is especially important, order the elements of a step, bar, or pie graph in a simple progression.

Adjust the range of values on the Y axis so that the visual impression reflects the appropriate information.

Although the values of the variables to be illustrated are determined by the numbers to be plotted, the range of the values of the population from which they are drawn is not. You may have data that range from 100 to 300, but the possible range is 0 to 500. In most cases, it is up to you to decide what range of values to include.

In which version, **don't** or **do**, are you more likely to notice that more blue-collar families were in the market for big-ticket items in 1991 than in 1990, but vice versa for white-collar families? Visually, the difference between the pairs of bars is more striking in **do** because the scale is not a uniform progression from zero. If it were, as it is in **don't**, the significant differences would be less apparent. According to the principle of compatibility, the visual impression produced by a display should convey actual differences and patterns in the data. To produce an accurate impression, display only the relevant range of the scale along the Y axis. Unless the zero value is inherently important, make the visible scale begin at a value slightly lower than the smallest value in the data, and the upper value slightly larger than the largest value. To show that you are leaving out a portion of the scale, make a clear gap in the Y axis (with two short slash lines or a zigzag, which will draw the reader's attention to this deletion) slightly above the origin. The gap in the axis signals a discontinuity in the scale along the Y axis. Adjusting the scale in this way allows readers to discriminate the important differences in the height of the content elements, which reflect the important differences in the data.

You may also want to consider excising the middle of the scale in the following situations: If you want the reader to see a difference in trends that is significant in the data; if you do not care about conveying the magnitude of the average differences between levels; or if the difference in trends would be difficult to see if the entire scale were presented because there is a large region of the scale in which there are no values plotted. Such excision will eliminate the unused portion of the scales, in essence producing two graphs, one above the other. This practice is unusual, however, so it is especially important that the slashes used to mark the break in the Y axis be easily detected.

don't

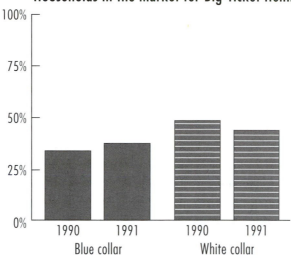

Households in the Market for Big-Ticket Items

do

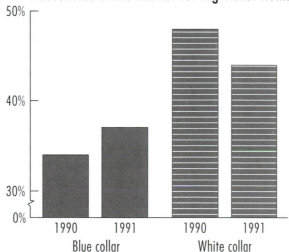

Households in the Market for Big-Ticket Items

Adjusting the scale along the Y axis makes the opposite trends for blue-collar and white-collar households immediately evident. A discontinuity in the scale can be indicated by slash marks or a zigzag in the axis.

Transform the scale if the resulting visual impression would more accurately convey the information.

Both these graphs present the same data, but they tell different stories. The top one obscures the differences among the methods. In contrast, the graph below reveals the key differences in the resolutions of the different methods. If in fact there were no real—that is, statistically significant—differences in the sensitivities of the methods, the bottom display would be an example of graphical deceit. When there really are significant differences, however, the principle of compatibility leads us to consider transforming the scale so that the visual impression reflects the actual pattern in the data. Is the range of measurements so great that important variations will be lost if a uniform scale is used? Do differences tend to increase with larger values? Is your purpose to illustrate trends or relative differences among similar cases, not the actual differences in overall amounts? If so, consider using a logarithmic scale on the X or Y axis, or both, as appropriate.

A logarithm is an exponent. When, as often, it is based on 10, it indicates the power to which 10 would have to be raised to produce a certain number. The log of 10 is 1, the log of 100 is 2, and so forth. The same actual distance on a logarithmic scale represents increasing amounts as one moves farther along the scale. Each equal increment represents multiplying the value of the previous increment by another factor of 10: If the first ticked-off space along the Y axis marked off values from 0 to 10, the next would specify values 10 to 100, and the next values 100 to 1,000. Thus, logarithmic scales compress differences among large numbers, because the same physical distance on the scale stands for increasingly large increments as the numbers get larger. Using a logarithmic scale is desirable if doing so makes important variations more discriminable, or if there is a great range of values on the axis and differences increase with larger values. If the scale is transformed, it is critical that it be labeled appropriately.

Maximal Resolution of Methods Used to Study the Brain

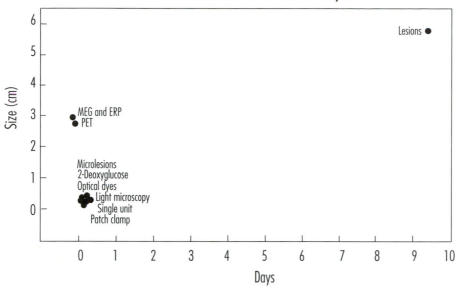

Maximal Resolution of Methods Used to Study the Brain

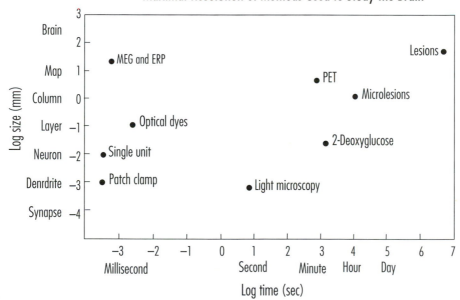

Using logarithmic scales here allows the reader to see differences among the higher-resolution techniques that otherwise are completely obscured.

● **recommendation** ───

Do not abbreviate or transform the scale on the Y axis in a stacked-bar graph.

The main point of a stacked-bar graph is to convey information about the relative proportions of different components of the whole. Excising part of the scale on the Y axis, as in **don't**, or using logarithms or another nonlinear transformation, will alter the visual impression so that the sizes of the segments no longer reflect the relative proportions of the components; the visual appearance will not be compatible with what is symbolized.

● **recommendation** ───

Arrange interval values on the X axis of a line graph at distances proportional to their magnitude.

The sharp increase in consulting revenues at Arthur Andersen, Inc., implied in **don't** did not in fact happen. Variations in the heights of lines suggest variations over continuous quantities when three or more points are placed along the X axis. The principle of compatibility implies that we should arrange the values at distances proportional to their actual values. If you are graphing number of ulcers for employees who have been on the job for one, two, and four years, the tick mark for year 4 should be placed at a distance from year 2 that is twice that between year 1 and year 2. Otherwise, the rate of rise and fall will not accurately reflect the trends in the data.

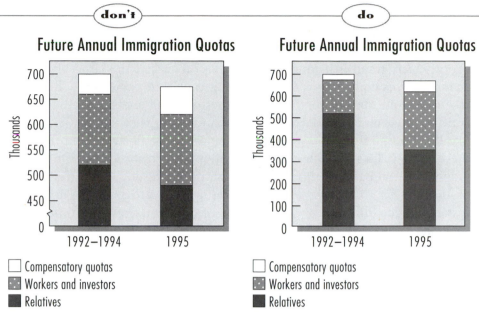

The sizes of segments are meant to reflect the relative proportions of components; if the scale is adjusted the visual impression is necessarily misleading.

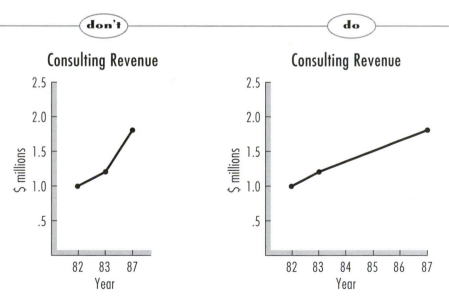

If values on an interval scale are not placed at correct proportional distances along the X axis, the content line will misrepresent trends in the data.

Graph positive and negative values relative to the zero baseline.

From which graph is it more obvious that Japan had the largest negative change in this composite index of leading economic indicators? In which case is it more obvious that the economies of Japan and the United States contracted, whereas those of Germany and Italy grew slightly? A graph like **do** is useful if there is a negative element to be portrayed; in this case there were years in which contraction, not growth, occurred. On such a graph the zero baseline should be indicated by a heavy line in the middle of the axis so that some bars hang below the zero origin. Our visual systems notice differences, and this type of display makes such differences explicit.

If specific values correspond to watershed points, it may be useful to include a reference line to indicate these points. For example, if you are graphing the average weight of a herd of cows for each day over a period of a year, you might want to insert vertical reference lines every two months, when the type of feed was changed. These lines should be clearly labeled on the X axis.

In a T-shaped framework, the X axis should be marked using the same scale on both sides of the Y axis.

The two segments of the horizontal line that is the X axis function as two scales for the dependent measure, one extending to the left of the vertical Y axis and one to the right. In keeping with the principle of compatibility, the same visual extent of bars on the two sides should signal the same amount, as in **do**. Similarly, to facilitate comparison of the bars in each pair, the left and right bars should have the same origin. If they do not, the visual impression may be misleading.

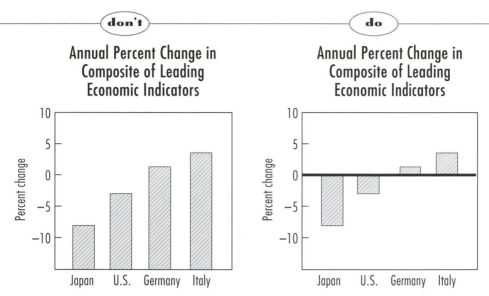

When positive and negative values are compared, make sure that the zero baseline is salient.

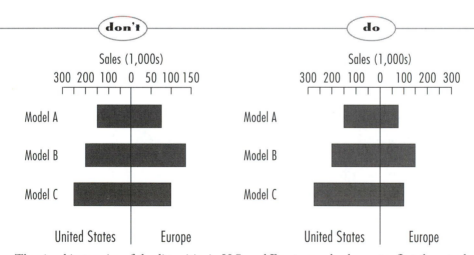

The visual impression of the disparities in U.S. and European sales does not reflect the actual disparities because different scales are used for each of the two regions.

Creating the Tick Marks

Tick marks indicate increments along a scale; they demarcate steps of increased amount (along an interval or ordinal scale) or the presence of a particular entity (along a nominal scale).

● **recommendation** ──●

Tick marks should extend inside the axis.

Having the tick extend inside the axis helps to group the label perceptually with the corresponding portion of the content.

● **recommendation** ──●

Place tick marks at regular intervals.

Remember that visible differences not only attract the eye but also lead the reader to expect additional information; if there is no such information forthcoming, do not signal it. One notable exception to this general recommendation sometimes occurs when a logarithmic scale is used; in that case, an increasingly dense placement of tick marks signals that the axis is calibrated in this scale.

● **recommendation** ──

Place larger ticks at labeled values and halfway between labeled values.

If the reader is meant to be able to estimate values that fall between labeled values, put a larger mark at the labeled value and at the value halfway between labeled values. For example, if the X axis labels intervals of ten years, place a heavy tick mark at each ten-year mark and a slightly less heavy one at the five-year increments, as in **do**. However, if intermediate point values are not important, make all the intermediate ticks the same size (informative changes and relevance are the key principles here).

● **recommendation** ──●

Do not place ticks or labels on the Y axis of a T-shaped framework.

In a T-shaped framework the Y axis is the central vertical line; it should not have ticks or any direct labels. In a horizontal bar graph, the labels are placed along the Y axis to promote grouping with the bars; in a graph with a T-shaped framework (a side-by-side graph), a second bar appears in the space that would otherwise be used for a label.

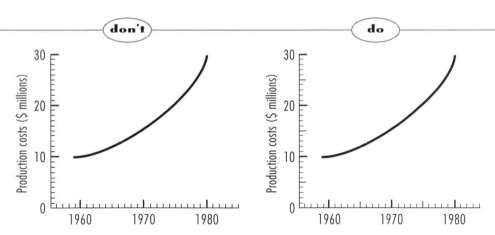

Intermediate heavier tick marks help the reader to estimate values along the content. An inner grid would also be necessary if the reader is meant to extract relatively precise values along the trend.

Use tick marks if specific values are important.

The principle of relevance suggests that tick marks should not be used if your aim is to portray a general trend, not to provide actual measurements. Both versions of the Microsoft graph are meant to show that the value of stock increased dramatically relative to the Dow Jones Software Index. For this purpose, the version on the right is preferable to that on the left because the minor fluctuations illustrated in the left-hand version are not pertinent and serve merely to distract the reader.

Creating the Labels

If the labels are too small, group improperly, or otherwise contravene principles of perception and cognition, the display may be uninterpretable. The following recommendations apply to all labels in a display.

Ensure that labels are easily detectable.

How often have you seen a display like **don't**? The designer apparently forgot that the display was going to be reduced! Ensure that labels and all other parts of the display are clearly discernible—and will remain so even if the size of the display is reduced before reaching the end-user. A quick check of the effects of reducing your display can be made by your computer graphics program or a copying machine with a reduction feature.

The most common violation of the principle of detectability involves size. Optimum size depends on two factors, the absolute size of the characters and the distance of the reader, both of which may vary (the distances at which a road sign, a blackboard, and a newspaper must be legible are very different). Thus, the "best" size for legibility is expressed as a relationship: the height of the characters divided by the distance from the reader. In studies, subjects were asked to move toward a display until they could just read it; measurements were taken at this point and the ratio calculated. The ratio is expressed in radians, the unit of the proportion of an arc of a circle (in effect, the height of the character) to the radius of that circle (the distance of the reader from the display). The U.S. military requires letters at least .007 radians (which corresponds to 24 minutes of arc), the proportion of size to distance at which people can read virtually perfectly.

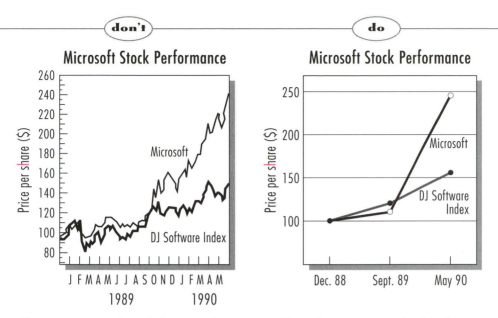

To portray a general trend, do not supply numerous tick marks or extraneous detail in the content.

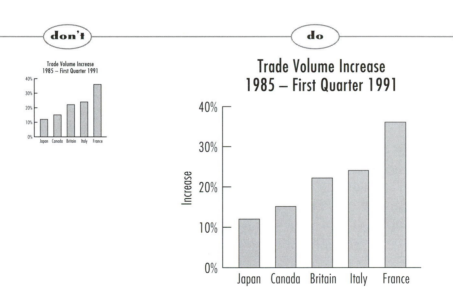

Plan ahead!

Avoid fonts in which letters share many features.

It is easier to distinguish a dog from a cat than a wolf from a German shepherd. The more characteristics that two objects have in common, the more information the visual system must register before it notes a difference. The principle of *discriminability* states that in order to be distinguished from one another, marks must differ by a minimal proportion. An addition of half an inch to a line is immediately noticeable if the line is only 1 inch long, but would be lost if the line were 6 feet long. Except for very large or very small starting levels, a constant proportion of the smaller value must be added in order for a larger value to be distinguishable. This principle of discriminability, or Weber's law, applies to size, lightness, thickness, density of dots, cross-hatching, and type of dashes, as we shall see in Chapter 7. Again, the mind is not a camera.

Fonts in which letters share many features are difficult to read because the reader must look carefully at each letter. For example, compare abcdefghijklmnop, ABCDEFGHIJKLMNOP, and *abcdefghijklmnop*. The letter forms in each of the last two fonts are more similar than in the first. Avoid using all upper-case letters, which are much more alike than a mixture of upper-case and lower-case letters. Also avoid italics if in the font you are using that style tends to make the letters alike.

Use visually simple fonts.

The desire to make an attractive display presents the temptation, succumbed to in **don't**, to use a "fancy" font. Such fonts are difficult to read and should be avoided. Use either serif type or sans serif type (this is serif type; this is sans serif), but be sure that the particular font is not so complex that the letters are hard to discern.

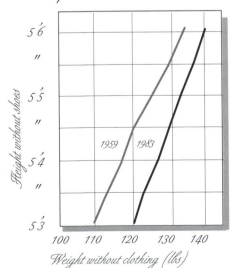

don't

Height and Weight (average) for Women, Medium Frame

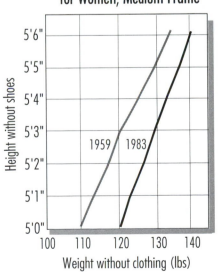

do

Height and Weight (average) for Women, Medium Frame

Select a font or style of lettering for its legibility.

Words in the same label should be close together and typographically similar.

Take advantage of the principles of perceptual organization to ensure that words in the same label cohere as a unit, as in **do**; they must not group individually with different parts of the display. The appropriate letters, numbers, and words should be relatively close to one another, and should be the same size, the same color, and the same brightness. Labels should not be so close to one another that they group improperly, as do the country labels in **don't**.

Use the same size and font for labels of corresponding components.

The labels for each of the same type of component (wedges, bars, etc.) should be the same size and font, as should the labels for the corresponding components of a multipanel display. (Similarly, labels on the axes of line and bar graphs should be the same size and font.) The principles of perceptual organization lead us to group similar forms into units, making a display so labeled easier to interpret at a glance. Furthermore, when we see a difference, we expect it to mean something (recall the principle of informative changes) and are confused if it doesn't.

The more salient labels should label the more general components of the display.

The principle of salience recognizes that the visual system reflexively attends first to extreme values and large differences in line length and width, shading, color, and other visual properties. We are led to notice heavier lines before thinner ones, and lines that contrast with the color of their backgrounds before lines of a similar color. Similarly, brighter colors are detected before dimmer ones, and thicker bars before thinner ones. Thus, the label for the display as a whole should be more salient than the labels for any parts, and labels for the individual panels in a multipanel display should be more salient than labels of wedges, objects, or segments.

don't do

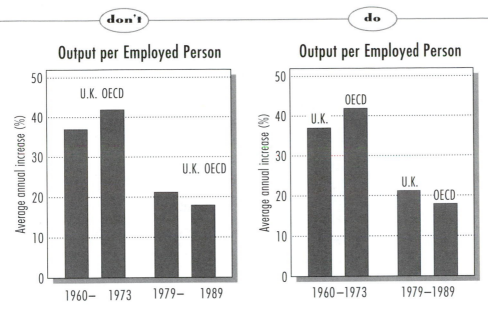

If labels group improperly, the reader must work to see what goes with what.

don't do

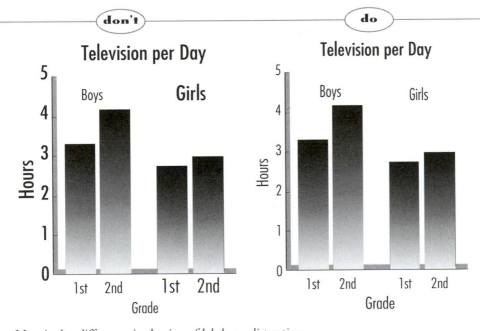

Meaningless differences in the sizes of labels are distracting.

• **recommendation**

Use the same terminology in labels and surrounding text.

Using different terms in a display and in the accompanying text suggests that they mean different things (recall the principle of informative changes). The same animals should not be labeled "birds" on the graph and referred to as "fowl" in the text.

• **recommendation**

Label a dimension with an "unloaded" term.

Many oppositional terms, such as high/low, near/far, above/below, light/dark, and so on, have a marked asymmetry in our use of the words. If you ask, "How high is it?" you are not implying that something is particularly high; but the question "How low is it?" implies that it is low. "High" (which is called the "neutral" or "unmarked" term) names the dimension as well as a pole, whereas "low" names only a pole of the dimension. "Low" is a loaded term—it strongly implies a value. Similarly, "near," "below," and "dark" are loaded terms. When you describe the two ends of the visual continuum, think about the words; is either loaded? Use the term that does not commit you to a particular end of the continuum.

• **recommendation**

Axis labels should be centered and parallel to their axes.

Compare **don't** and **do**: In which is it clearer what the axes represent? If labels are set centered and parallel to their axes, the principle of common fate will group the label and axis together, as occurs in **do**. If this recommendation cannot be followed because of technical limitations, such as a poorly designed computer graphics program, make sure that the label for an axis is closer to that axis than to the other axis. Avoid putting labels in the lower left, where perceptually they could be grouped with either axis.

don't   do

Career Preferences

Rated popularity

Industry Farming   Lumber   Mining

Career Preferences

Farming   Lumber   Mining
Industry

Axis labels that do not group clearly with their respective axes are ambiguous and require time to decipher.

Value labels should be centered close to their tick marks.

Each label of a value along the Y axis—numbers counting off the barrels of oil, tons of peanuts, etc.—should be centered to the left of the appropriate tick mark and closer to it than to anything else, so the label will be associated with the tick. Similarly, each label of a value along the X axis should be centered beneath the appropriate tick mark, and closer to it than to anything else. These guidelines are followed in **do**.

Label ticks with round numbers and regular intervals.

The principles of cultural convention, relevance, and informative changes all lead me to recommend round numbers and regular intervals, as in **do**, for tick marks along the axes: 5, 10, 15, 20—not 5.3, 10.4, 15.5, 20.5. The only exception to this rule occurs when a specific value has a special meaning (such as 98.6 degrees Fahrenheit). Some computer graphing programs do not respect this recommendation; my advice is not to buy those programs.

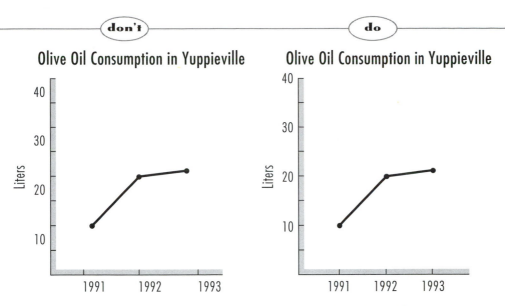

If labels are not immediately next to the appropriate tick mark, the reader will have to pause to determine the association.

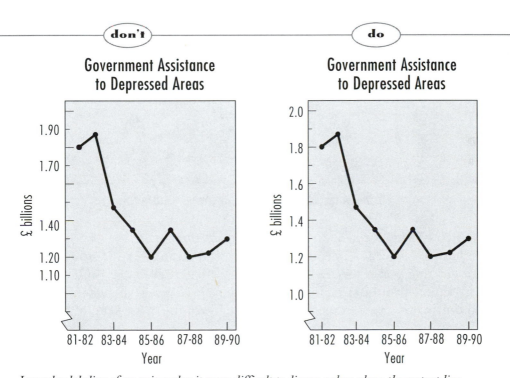

Irregular labeling of an axis makes it more difficult to discern values along the content line.

● **recommendation** ————————————————————————————————

In T-shaped frameworks, label the left and right sets of bars.

This graph, which shows relative percentages of votes for the British Labour (on the left, of course) and Conservative (on the right) parties, is a good use of a side-by-side format. The left and right sets of bars correspond to different levels of an independent variable. Put labels over or under each side, as space permits.

● **recommendation** ————————————————————————————————

Do not use redundant labels on the axes of T-shaped frameworks.

If, as is usually the case, the left and right portions of the Y axis delineate the same dependent measure, use one label for both halves. This practice not only reduces clutter but also is consistent with the principle of relevance.

● **recommendation** ————————————————————————————————

Label each component of the content material.

To be easily interpreted, each information-bearing aspect of a display should be labeled, as in **do**. This recommendation follows from the principles of relevance and informative changes. Recall that part of the problem with the "Nutritional Information" display on page 12 is that there are no labels on the arrows that join the center visual table to the other panels.

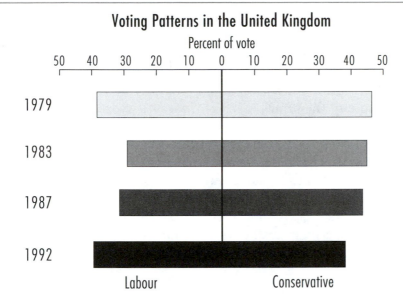

Voting Patterns in the United Kingdom
Percent of vote

Labels for the left and right sides can be at the top or bottom, wherever there is room—provided that proximity properly groups the labels with the content bars.

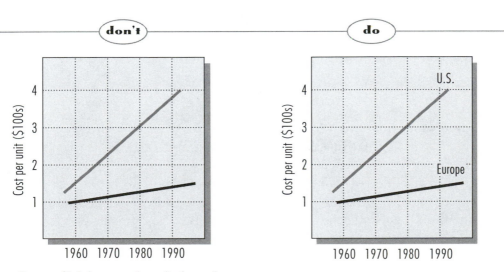

Absence of labels can render a display useless.

• recommendation ──────────────────────────────────

Label content elements directly.

If possible, place labels within wedges, objects, or segments. If this is not possible because the labels would be too small to be read, then place them as close as possible to the corresponding component, as in **do**, so that they group (because of the principle of proximity). If even this is not possible, or creates clutter, then, and only then, use a key.

Although I have stressed placing labels beneath the X axis in bar graphs or step graphs, labels can instead sometimes be placed above the bars or steps. If the bars or steps are not arranged in a progression, however, it can be difficult to locate labels above them, and thus I recommend putting the labels beneath the X axis in the usual way. If the tops of the content elements are easy to track, as in the step graph on page 77, then labels are effective when placed directly on the content elements.

Creating the Title

The same guidelines for creating the title apply whether the graph stands alone or is part of a multipanel display.

• recommendation ──────────────────────────────────

The title should indicate what questions are answered by the graph.

A graph should be designed to help the reader answer a specific set of questions, and the title should be a signpost to those questions. Which title is more helpful, the one for **don't** or for **do**? When formulating the title, ask yourself: What material is illustrated? Does it apply to a particular time and place? Include only the relevant information in the title.

• recommendation ──────────────────────────────────

The title should be typographically distinct.

A larger or different font should be used for the title to help to set it off (exploiting the principles of perceptual organization); furthermore, this font should be sufficiently distinct to catch the eye immediately, making the title the most salient element of the display.

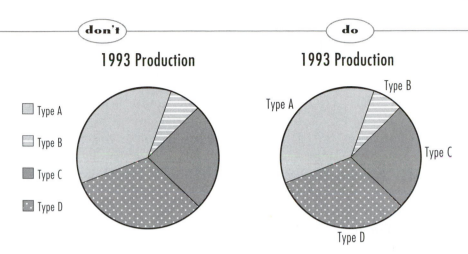

don't　**do**

Direct labeling greatly reduces the effort required to read a display.

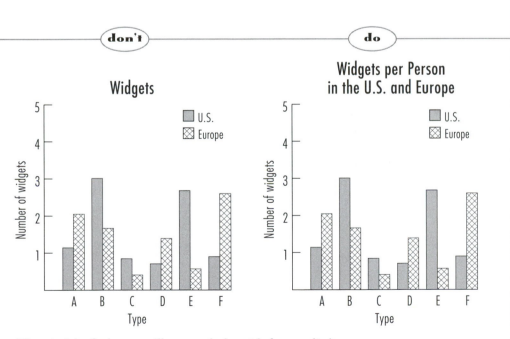

don't　**do**

The principle of relevance will suggest the best title for your display.

The title should be centered over or under the display.

Convention leads the reader to expect the title in one of these two places. If content labels or elements appear at the bottom of the display, it is better to put the title at the top. If you put the title at the bottom, it may be close enough to the other labels or elements to be improperly grouped. If the title is off center, it may be grouped with an axis, as in **don't**. If, however, the display appears in a book or lengthy article with a distinctive but homogeneous design (for example, one in which the titles are always in the left margin), the reader will grow familiar with this design and will group the title properly. Note that such departures from the norm initially require the reader to work harder than do the standard arrangements.

The Next Step

If you chose in Chapter 2 to make a pie graph, divided-bar graph, or visual table, you have used this chapter only to create the title and content labels. Your display is essentially finished, and you should now check Chapter 8 to be sure you have not created a misleading graph. If your choice has an L- or T-shaped framework, you have created that framework from the recommendations in this chapter and are ready to create the content elements. Depending on the type of graph format you have chosen, turn to Chapter 4, 5, or 6.

don't

do

Remains of World Reserves
(Based on 2.3% Exponential Growth Rate)

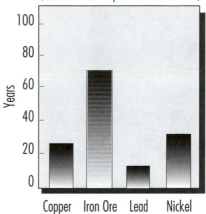

Remains of World Reserves
(Based on 2.3% Exponential Growth Rate)

The title should clearly be seen to apply to the display as a whole, not an axis or other component.

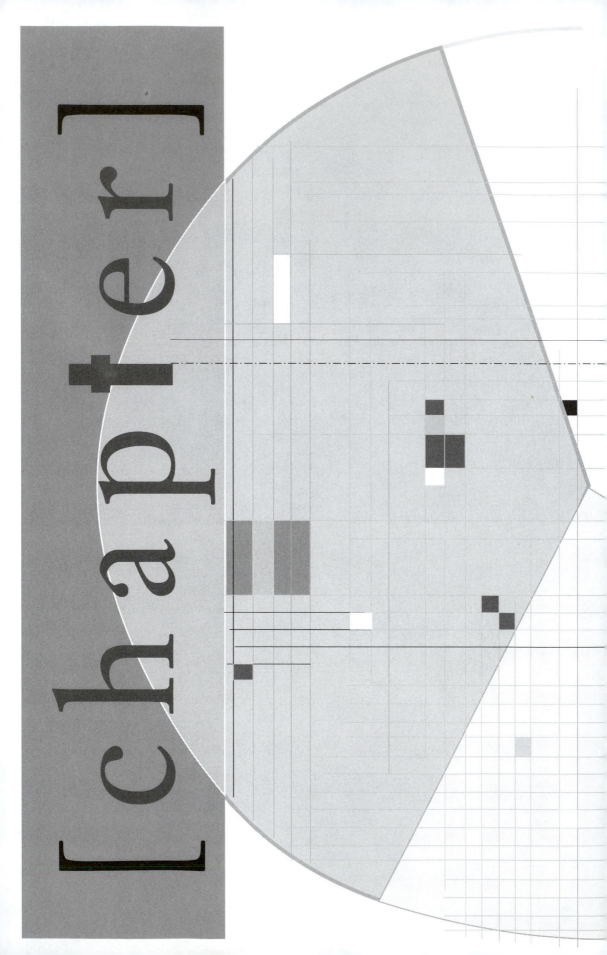

[chapter]

Creating Pie Graphs, Divided-Bar Graphs, and Visual Tables

4

Pie graphs, visual tables, and divided-bar graphs are the simplest displays that convey relations among quantities. This chapter provides detailed advice for creating the content elements of these graphs. If you want to include shading, hatching, color, three-dimensional effects, inner grid lines, background elements, a key, or a caption, turn to Chapter 7 after you have finished with the material in this chapter. Recommendations for creating labels of the content elements are summarized in this chapter; the reader should turn to Chapter 3 for the reasons for these recommendations, and for recommendations for creating the title.

Pie Graphs

The content elements of pie graphs are wedges, larger wedges signaling greater amounts. After you have drawn the wedges, you will probably want to fill them in with different shading, hatching, or color, to make each distinct; Chapter 7 provides recommendations for these features of displays.

● **recommendation** ──

Draw radii from the center of a circle.

The straight lines that define wedges should originate at the precise center of a circle, as in **do**, not at a point off center. The proportion of the depicted amount is reflected by the size of the angle of the wedge. In a graph illustrating the proportions of workers of different ages, if half of all workers are in the 20–30 age range, then that wedge should be 180 degrees—half of the circle; if one quarter of the workers are in the 30–40 age range, then that wedge should be 90 degrees—a quarter of the circle. And so forth.

● **recommendation** ──

Explode a maximum of 25% of the wedges.

If you decided to use an exploded pie, you must determine which part or parts to emphasize. The visual system is sensitive to changes in a stimulus. If too many wedges are exploded, as they are in **don't**, they will not be emphasized because the principle of saliency is being violated. I offer 25% as a rough guideline; there is no hard and fast percentage. The critical consideration is that enough wedges remain in the pie to make the exploded wedges disrupt the contour of the whole.

● **recommendation** ──

Arrange wedges in a simple progression.

Unless there are reasons to order the wedges in a specific way, it will be easiest to compare the wedges if they are arranged in order of size; the principle of cultural convention suggests that the smallest wedge should be at the top, with size increasing in a clockwise progression. Similar-sized wedges will then be next to each other, as they are in **do**, facilitating comparison. Also, the visual system can "chunk" a progression into a unit, making it easier to remember than a set of arbitrarily arranged elements and thus expanding our limited short-term memory capacity.

don't   **do**

Factory Workers by Age Group

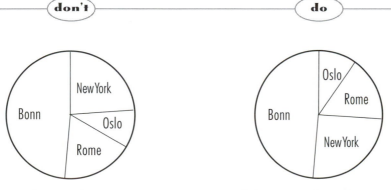

Factory Workers by Age Group

Angles, arcs, or chords reflect relative area only if the wedges meet at the center of the circle.

don't   **do**

If too many wedges are exploded, none stands out.

don't   **do**

In Western culture, readers expect quantities to increase clockwise around a circle.

● recommendation ───●

Follow the guidelines for labels given in Chapter 3.

Ensure that labels are easily detectable; avoid fonts in which letters share many features; use visually simple fonts; words in the same label should be close together and typographically similar; the more salient labels should label the more general components of the display; use the same size and font for labels of corresponding components; use the same terminology in labels and surrounding text; label each component of the content material; label content elements directly. These recommendations are illustrated and discussed in detail on pages 88–101.

● recommendation ───

Place labels in wedges provided that they can be easily read.

The principles of perceptual organization lead to our grouping objects that are physically juxtaposed. Placing labels within the wedges, as in **do**, will clearly associate them with the appropriate wedge. However, do not put labels in wedges if the labels would be so small that they would be difficult to read.

● recommendation ───

Place labels next to wedges if they cannot all fit within wedges.

If you cannot fit *all* the labels in wedges, then place *all* of them next to the appropriate wedges; do not put some labels in wedges, and others next to wedges, as in **don't**. The reader will expect any change to convey information, and will otherwise expect consistency (recall the principle of informative changes).

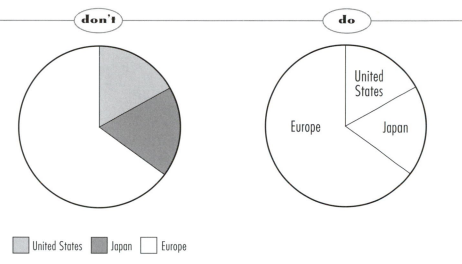

don't　　　**do**

Labels group best with content elements if they are actually placed within the elements.

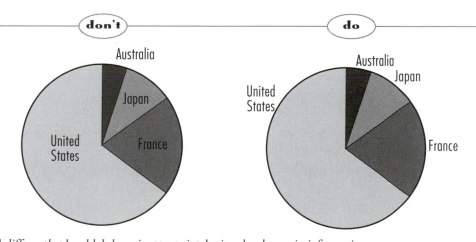

don't　　　**do**

A differently placed label can inappropriately signal a change in information.

Divided-Bar Graphs

A divided-bar graph is created by drawing a rectangle—usually vertically oriented—and dividing it into segments.

• **recommendation** ————————————————————————————————————

Make the bar wide enough to demarcate segments clearly.

Because the information in this format is conveyed by the relative sizes of the segments, the bar must be wide enough that the individual segments are seen clearly, as in **do**. Moreover, it is best to make the segments wide enough that labels can fit directly within them, promoting better perceptual grouping.

• **recommendation** ————————————————————————————————————

If you use a scale, place it on the left border.

If you include a scale (always 0–100%), mark it directly on the left border of the bar itself, as in **do**. The tick marks should be created and labeled in accordance with the recommendations in Chapter 3 (pages 84–88 and 96–97).

If you have more than one bar in your display, however, set the scale *apart* from the bars, at the left. If it were placed directly on the border of the first bar, it would be perceptually grouped with that bar only.

• **recommendation** ————————————————————————————————————

Draw segments with parallel horizontal lines.

The segments should be divided by parallel horizontal lines, as in **do**. If the lines are not parallel, the reader will think that the quantities are varying with time or another variable along an X axis—the divided bar will be mistaken for a layer graph. According to the principle of informative changes, every change in a display will be taken to convey information; the height along the top of a segment should be constant.

110

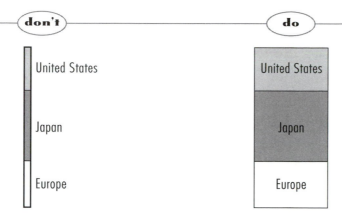

Use a bar that is wide enough to ensure that the regions are distinct and to allow the labels to be placed within the content elements.

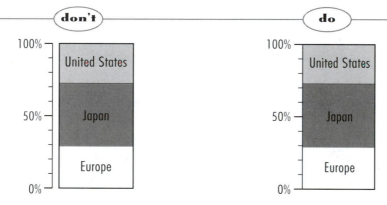

A scale on the left border of the bar facilitates lining up tick marks and content segments, but if there is more than one bar a separate scale is preferable, so it does not group with one bar only.

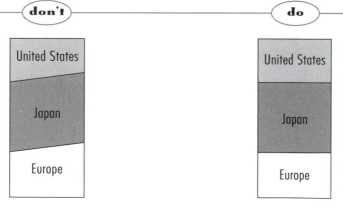

Segments drawn with slanting lines improperly suggest a layer graph.

In multiple divided bars, put segments that change the least amount at the bottom.

When designing a divided-bar graph to illustrate two or more independent variables, make it as easy as possible for the reader to compare the changes in the components. Look at **don't**: Are tape sales or singles sales increasing at a faster rate? Now look at **do**: Here it is relatively easy to see that tape sales are increasing at a greater clip. By putting the segment that changes the least amount at the bottom, the comparisons of changing heights will be relatively easy (in order to know the proportion, the reader must visually subtract the bottom of a segment from the top, which is easier when the baselines are similar).

Visual Tables

Visual tables are the least structured type of graph; all that is necessary is that the sizes of content elements reflect relative quantities. Such displays are often interesting because pictures of objects are used as content elements.

The appearance of the pattern should be compatible with what it symbolizes.

To respect the principle of compatibility, the appearance of a pattern should not conflict with what it symbolizes. Our visual systems register the properties of a stimulus at the same time that we comprehend what the stimulus symbolizes; if the properties of the pattern are not compatible with its meaning, those properties will interfere with our reading the display (the mind judges a book by its cover). As a consequence, we automatically perceive a larger region, a greater extent, a lighter area (on a dark background), or darker area (on a light background) as corresponding to a greater amount: *More is more*. Consider the amounts that were measured for each case, and ensure that this principle is not violated, as it is in **don't**. Further, the visual impression of a difference should correspond to the actual amount of the difference between the represented substances.

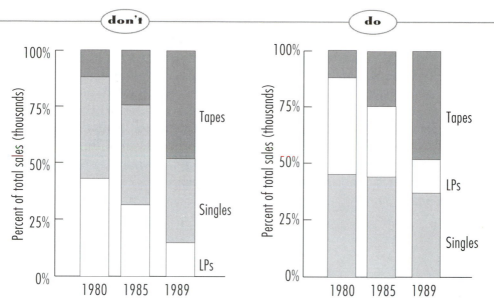

Putting "Singles," which varies least, at the bottom of each bar allows the reader to make comparisons among the three categories most easily.

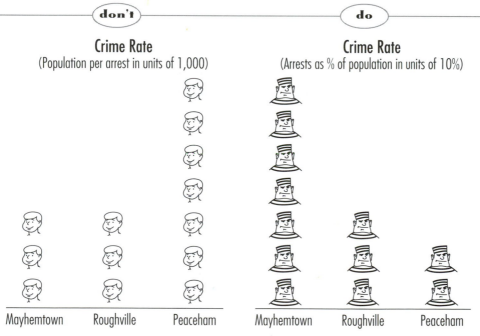

Do not convey "more" by "less." A greater extent for a smaller amount violates the reader's expectations.

───

Pictures used as content elements should illustrate the variables.

The principle of compatibility implies that pictures used as content elements either should depict the entities being represented, as in **do**, or should have conventional symbolic interpretations that label those entities (such as an icon of an envelope for a post office or a donkey for the Democratic party). If the amount of lumber produced in different regions is illustrated, pictures of trees of different heights would be appropriate, whereas pictures of objects made from trees, such as ladders or park benches, would not.

───

Compare extents at the same orientation.

The **don't** version immediately demonstrates why this recommendation is a good idea: It is much harder to derive a relationship of quantities among the three elements when there is no common baseline and we must assess the absolute value of each in isolation, remember it, and compare it with the others.

Extent is difficult enough to portray in a way we will perceive accurately: Our visual systems do not always tell the truth, and they systematically misinform us that a vertical line is slightly longer than a horizontal one of equal length. The famous top-hat illusion illustrates this distortion nicely—the hat is actually as wide as it is high. It has been shown that people tend to overestimate the length of vertical bars.

However, we do not systematically underestimate or overestimate differences in lengths at the same orientation. Thus, as long as you stick with a single direction, the lengths of objects in a visual table can convey relative quantities accurately.

don't

Trees in Yuppieville Park

do

Trees in Yuppieville Park

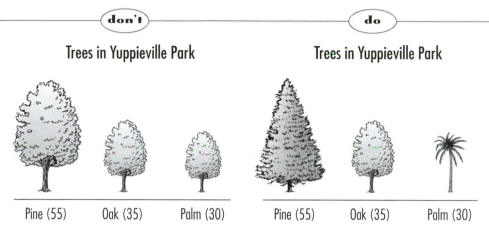

| Pine (55) | Oak (35) | Palm (30) | Pine (55) | Oak (35) | Palm (30) |

Pictures in a visual table should provide another kind of label, and should not interfere with the interpretation of the display.

don't

do

U.S. Exports of Manufactured Goods

U.S. Exports of Manufactured Goods

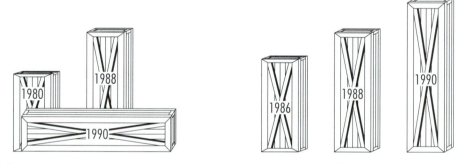

It is much easier to compare extents at the same orientation than at different orientations.

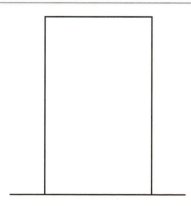

The top-hat illusion. Measure the height and the brim of the hat to convince yourself that their extents are the same.

Do not vary height and width to specify separate variables.

A designer might use rectangular objects to present seasonal totals for new accounts opened at a bank. The height of each rectangle would indicate the number of accounts that were opened during that particular season; the width, the average amount deposited in such accounts; and the third variable, the area, would indicate the total revenue for each season. This scheme is shown in **don't**. One problem with such a display is that we see a rectangle—not height, width, and area as distinct elements. There are two ways that we perceive combinations of visual properties. Separable properties can be registered individually; a reader can pay attention to one variable and ignore the other. The color and length of a line are separable, as are the length and orientation of a line. But some properties are automatically combined by the visual system, and the principle of *integral dimensions* states that these properties cannot be seen separately except with difficulty. The height and width of a rectangle are examples of integral dimensions: You cannot pay attention to the height without also registering the width and the resulting area—and we do not perceive differences in area very accurately (see page 24). The mind is not a camera; it actively organizes and interprets information. As a general rule, do not vary both the height and width of an enclosed area to represent changes in differing quantities.

The Next Step

Consult the recommendations in Chapter 7 for using shading, hatching, color, three-dimensional effects, background elements, and keys. Turn to Chapter 3 to create the labels for the individual content elements and the display as a whole.

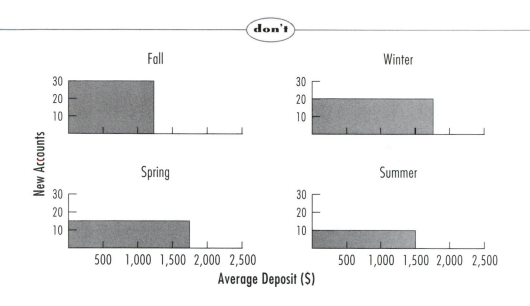

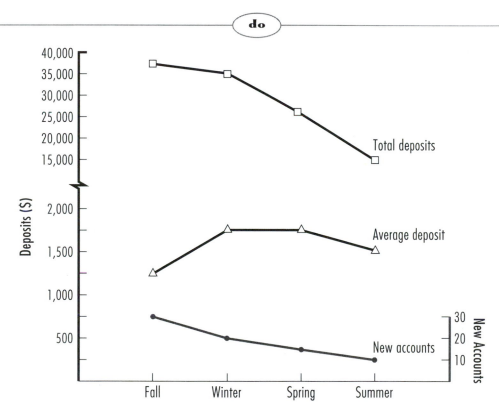

The line graph clearly indicates the number of new accounts, the average deposit, and the total deposits per season, as well as the relations among these measures. The same information is embedded in the four panels above, but considerable effort is required to decode it—we perceive height and width integrally.

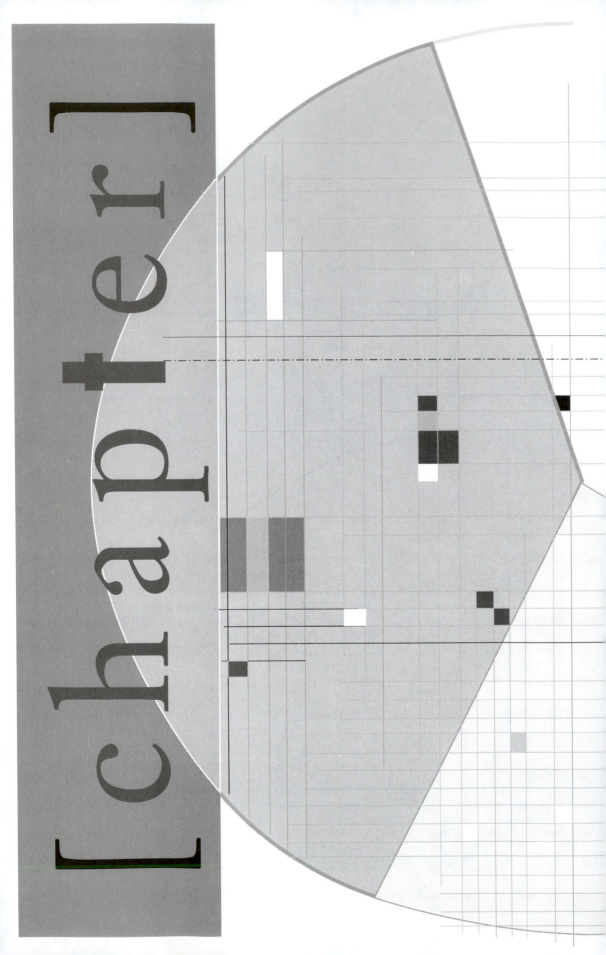

[chapter]

Creating Bar-Graph Variants

5

Bar graphs and their variants are used primarily to answer questions about quantities: how they compare and how they change. These graphs include a framework as well as content and labels. Create the framework and its labels using the recommendations in Chapter 3. Then create the content and its labels using this chapter. If shading, hatching, color, three-dimensional effects, an inner grid, a key, background elements, a caption, and/or more than one panel is needed, consult the recommendations in Chapter 7 after finishing this chapter.

Bar Graphs

A bar is any content shape whose extent (height in vertical bar graphs, length in horizontal bar graphs) is used to indicate an amount; rows of pictures—houses, trees, people, or other images—can be used as bars within a framework.

● recommendation

Corresponding bars should be marked in the same way.

Try to compare the performances of the two brands of Coca-Cola in 1989 from the **don't** graph. This is not as easy as it should be because in a misplaced attempt at variety in the way the bars are shaded, the designer used different shadings for corresponding elements. When more than one independent variable appears in a bar graph (as, here, brand and year), use different colors, shades, and/or different dashed, striped, or hatching patterns to indicate the bars for different variables. The corresponding bars in each cluster should be marked the same way so that they are grouped appropriately (exploiting the principle of similarity), as in **do**.

● recommendation

Arrange corresponding bars in the same way.

Arrange bars that depict the levels of a parameter in the same way in each cluster on the X axis. Say you want to graph numbers of young and old men and women voters for each of two counties. County would be on the X axis, and four bars, one for each of the four demographic groups, would sit over each county. The demographic groups—young men, old men, young women, old women—could be indicated by different texture. The order of the bars for voter group should be the same for each county, as in **do**. If it isn't, the principle of informative changes implies that the reader will waste time trying to make comparisons and wondering why the order differs.

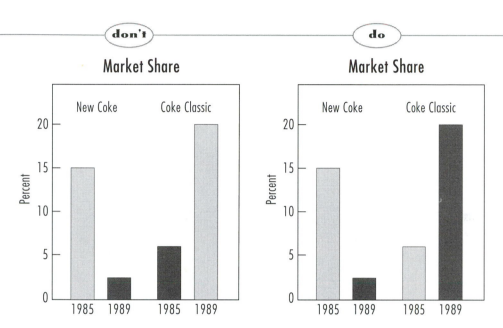

Inconsistent marking groups bars improperly, making it harder to compare corresponding elements.

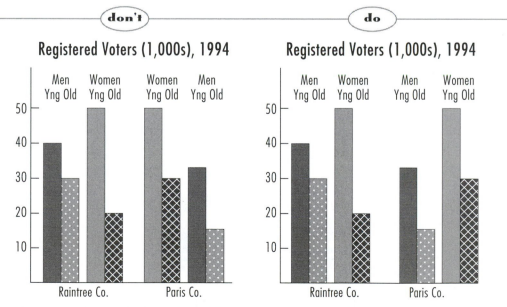

A gratuitous change of order makes it difficult to compare corresponding bars.

Occluded bars should not look like stacked bars.

Some bar graphs show two parameters by partially occluding one of the two bars in each pair, so that one appears to be standing in front of the other. Make sure that the "nearer" bar does not cover all or nearly all of the lower part of the "farther" one. If it does, your display will be misinterpreted as a stacked-bar graph, and your reader confused. At first glance, **don't** seems to indicate the *sum* of the percentage of public enterprises in the total economy for 1982 and 1986. In fact, as is evident in **do**, the two bars in each pair specify entirely separate measures.

Do not vary the salience of individual bars arbitrarily.

Unless the emphasis is intentional, no bar should stand out from the others as happens in **do**; if it does, the difference in salience will lead the reader to notice it first, and assume that it is more important than the other bars in the display.

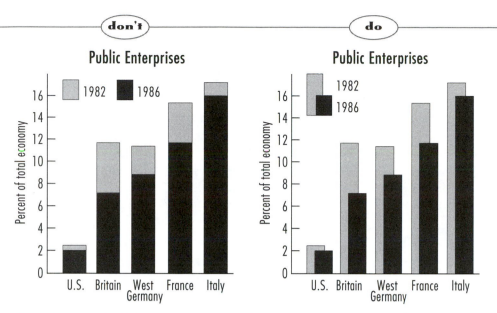

don't   **do**

Partial occlusion should leave no doubt that the display is not a stacked-bar graph.

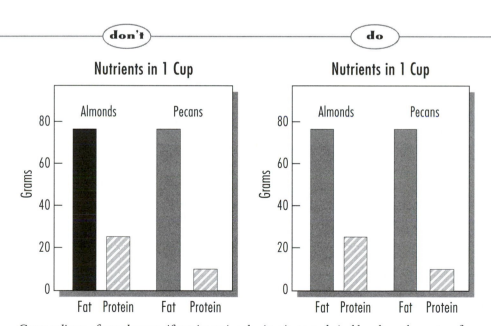

don't   **do**

Great salience of one element, if not intentional, gives it an undesirable role as the center of attention.

● **recommendation**

Leave space between bar clusters.

Is it easier to compare the market shares in 1985 for the three brands in **don't** or **do**? When two or more independent variables are included in a bar graph, group the bar clusters over each appropriate location on the X axis and leave extra space between the clusters. As a rule of thumb, the space between clusters should be about as wide as two bars. The principle of proximity will group the bars within a cluster but separate those in different clusters, as can be seen in **do**.

● **recommendation**

Bars generally should not extend beyond the end of the scale.

Bars should not extend over the top of the Y axis as in **don't** (or to the right of the X axis, in a horizontal bar graph) if the viewer is supposed to be able to extract specific point values; mentally continuing the axis and its scale requires effort, and the principle of relevance requires that necessary information be supplied. However, if the point of the display is not to show specific values but only to indicate that, say, housing prices have gone through the roof, you should not present extraneous details (such as tick marks, labels on the Y axis, and so on). Some of the most effective graphs in *Time* magazine have tall bars that extend into the text, driving home the point that some trend has exploded beyond its usual boundaries.

● **recommendation**

Use half- "I" error bars.

Error bars specify a range of measurement error around an average. For example, in a graph of the results of an opinion poll, error bars would indicate the range over which one could expect to find the average if a different sample had been polled at the same time. In a bar graph, show only the half of the "I" that extends above the bar, as illustrated in **do**. The error bars should be drawn in lines of the same thickness as those used for the framework, which will help prevent them from being overshadowed by more salient elements of the display.

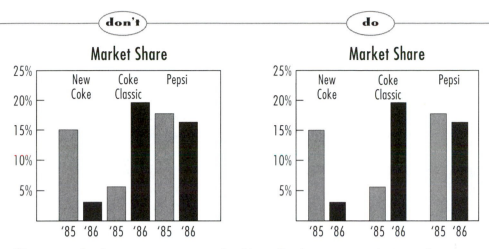

Six perceptual units are too many to apprehend immediately; proper grouping not only makes the display easier to take in but also helps to group the bars with their labels.

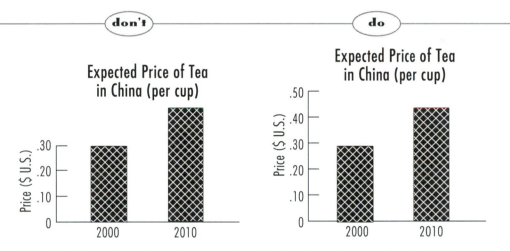

If the Y axis is too short, the reader cannot immediately estimate the price of tea in 2010.

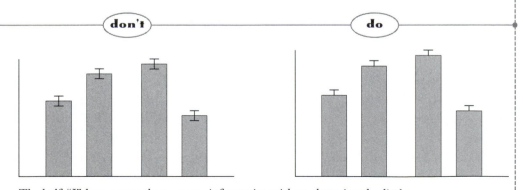

The half-"I" bars convey the necessary information without cluttering the display.

Use hierarchical labeling.

Which town had greater sales of single-family homes in 1991, Hillsborough or Pinellas? This is relatively easy to determine from **do**, but much harder from **don't**. When there are two or more independent variables (and so there is at least one parameter), use a hierarchical labeling system, one in which some labels are larger or bolder—that is, more important—than others. Such a scheme, illustrated in **do**, eliminates the need for a key, which would tax our short-term memory capacities. It also avoids redundant labeling, which clutters a display (thereby making it more difficult to discriminate the other elements), and specifying each of the relevant dimensions separately helps the reader to make specific comparisons.

By using labels of greater salience to label larger portions of the display, you draw the reader's attention to major labels before minor ones; in order to understand the scale values, the reader must know what variable is being graphed on that scale, and in order to understand *why* those variables are presented, the reader must understand *what* is being presented.

Follow the guidelines for labels given in Chapter 3.

Ensure that labels are easily detectable; avoid fonts in which letters share many features; use visually simple fonts; words in the same label should be close together and typographically similar; the more salient labels should label the more general components of the display; use the same size and font for labels of corresponding components; use the same terminology in labels and surrounding text; label each component of the content material; label content elements directly. These recommendations are illustrated and discussed in detail on pages 88–101.

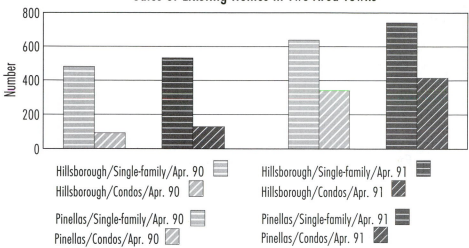

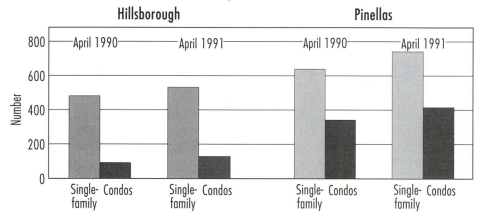

The bars are easier to compare along specific dimensions (dwelling type, town, year) when the labels specify the dimensions separately.

Stacked-Bar Graphs

Stacked-bar graphs are a hybrid of bar graphs and divided-bar graphs. They are like bar graphs in that the bars appear in an L-shaped framework, and they are like divided-bar graphs in that they display components of a whole. However, in stacked-bar graphs, the components do not sum to a constant-sized whole.

● **recommendation**

Put the segments that change the least amount at the bottom.

Putting the segment that changes the least amount at the bottom, as in **do**, makes it relatively easy to compare the changing amounts of specific segments. To read a stacked-bar graph, the reader must visually subtract the value at the bottom of a segment from the value at the top to see the amount, a feat that is more easily accomplished if the bottoms are as nearly at the same heights as the data allow.

● **recommendation**

Follow the guidelines for labels given in Chapter 3.

Ensure that labels are easily detectable; avoid fonts in which letters share many features; use visually simple fonts; words in the same label should be close together and typographically similar; the more salient labels should label the more general components of the display; use the same size and font for labels of corresponding components; use the same terminology in labels and surrounding text; label each component of the content material; label content elements directly. These recommendations are illustrated and discussed in detail on pages 88–101.

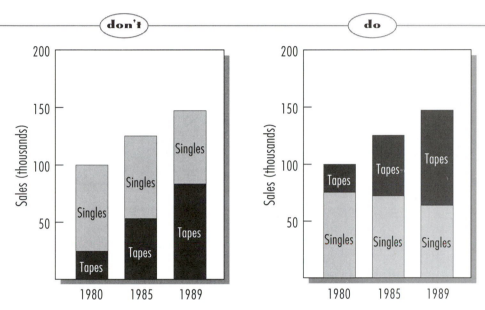

Because "Singles" changes relatively little, putting it at the bottom provides a relatively constant baseline for "Tapes," which helps the reader to compare the two categories.

Step Graphs

Step graphs are bar graphs in which the bars are contiguous and the internal borders have been removed, resulting in a step pattern. The content element of a step graph is a line—the outline of the steps—that varies in height in discrete increments. This format is best for displaying a single independent variable. Because there is only one independent variable, which is labeled directly or along the X axis, there are no additional labels for the content material.

● **recommendation** ———————————————————————————————————

Ensure that the line is detectable.

The content line should be heavy enough to be noticed at a glance, as in **do**. The content lines should be at least as heavy as the framework; indeed, most of the graphs in this book—the **do** versions, that is—show content lines three times heavier than the framework.

● **recommendation** ———————————————————————————————————

Make the steps of equal width.

In which version of the graph does U.S. investment in Germany look more important? Because of the principle of integral dimensions, the reader will perceive a wider step as representing a greater amount, even if that step is the same height as another, narrower step. Make sure the steps are of equal width, not drawn as in **don't**.

130

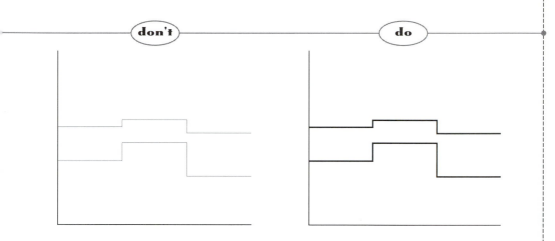

The reader should not have to search for the content of a graph.

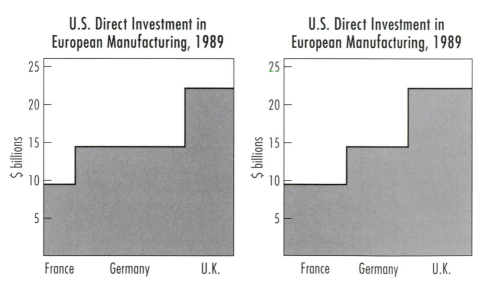

Making one step wider than the others—even if its height is the same—gives it unequal importance.

Fill the area under the line with a single pattern or color.

The primary virtue of step graphs is that they produce patterns that indicate specific trends. As is evident in **do**, if the area under the line is filled with a color or texture that is different from the background, the principle of similarity will lead the reader to see that area as a single shape, facilitating recognition of a trend. Only a single color or texture should be used; otherwise, the impression of a single region will be disrupted.

If the area under the line is filled with a color or texture different from that of the background, the salience of the content will be increased and the eye will be drawn to this visual difference. If the reader is likely to be familiar with the topic of the display (say, from the surrounding text), this is fine. However, in some situations you may want to ensure that the reader examines other components of the display, such as the title, carefully; in these cases consider whether the content material should be more salient than the framework and labels.

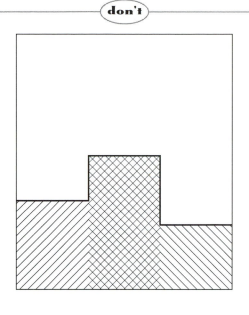

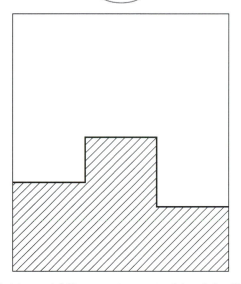

The shape of a series of rising and falling steps is meaningful and should not be disrupted.

Side-by-Side Graphs

In many respects, side-by-side graphs are like horizontal bar graphs; they differ in the type of framework used.

Align the starting points of corresponding bars.

The power of side-by-side graphs arises from the perceptual units that are created by pairs of corresponding bars. Therefore, the bars in each pair must be aligned so that the same extent corresponds to the same amount, as in **do**. The central vertical axis can produce an asymmetric pattern, allowing the reader to see the inequalities in the bars immediately.

Order the bars to emphasize interactions.

The visual system is sensitive to differences in patterns. Thus, if the Y axis is a nominal scale and the levels have no inherent order, arrange the left series of bars to form a regular progression, as in **do**; the differences between those bars and the ones on the right will then be readily apparent.

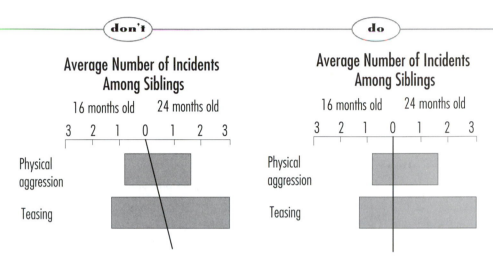

**Average Number of Incidents
Among Siblings**

16 months old   24 months old

3 2 1 0 1 2 3

Physical
aggression

Teasing

**Average Number of Incidents
Among Siblings**

16 months old   24 months old

3 2 1 0 1 2 3

Physical
aggression

Teasing

The slanted Y axis shifts the starting points of the bars, resulting in a distorted impression of their relative magnitudes.

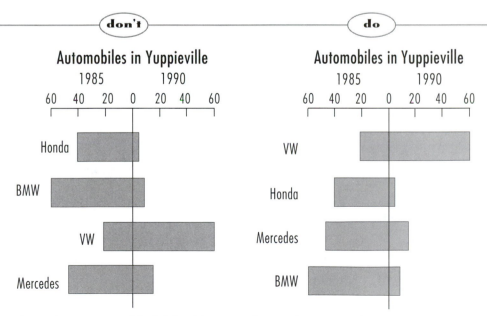

Automobiles in Yuppieville

1985   1990

60 40 20 0 20 40 60

Honda

BMW

VW

Mercedes

Automobiles in Yuppieville

1985   1990

60 40 20 0 20 40 60

VW

Honda

Mercedes

BMW

A smooth progression of the left-hand bars provides a good contrast to the pattern of the right-hand bars.

● **recommendation** ───●

Follow the recommendations for labels given in Chapter 3.

Ensure that labels are easily detectable; avoid fonts in which letters share many features; use visually simple fonts; words in the same label should be close together and typographically similar; the more salient labels should label the more general components of the display; use the same size and font for labels of corresponding components; use the same terminology in labels and surrounding text; label each component of the content material; label content elements directly. These recommendations are illustrated and discussed in detail on pages 88–101.

● **recommendation** ───

Label pairs of bars consistently on one side.

A label for each pair of bars should be placed either to the left or right of the pair, close enough to be grouped both by good continuation and proximity. The principle of informative changes implies that the labels should be in the same places for each pair. As a general rule, place labels on the left, following the Western convention of reading left to right.

The Next Step

Now that you have created your graph in outline, the next step is to fill in the bars with shading, hatching or color. You may need to add a key. Other possibilities are three-dimensional effects, an inner grid, a caption, background elements, and a multipanel display. Turn to Chapter 7, where you will find recommendations for creating all these features.

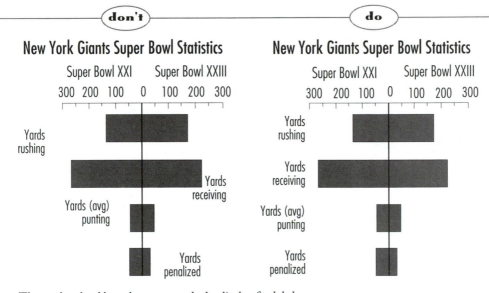

The reader should not have to search the display for labels.

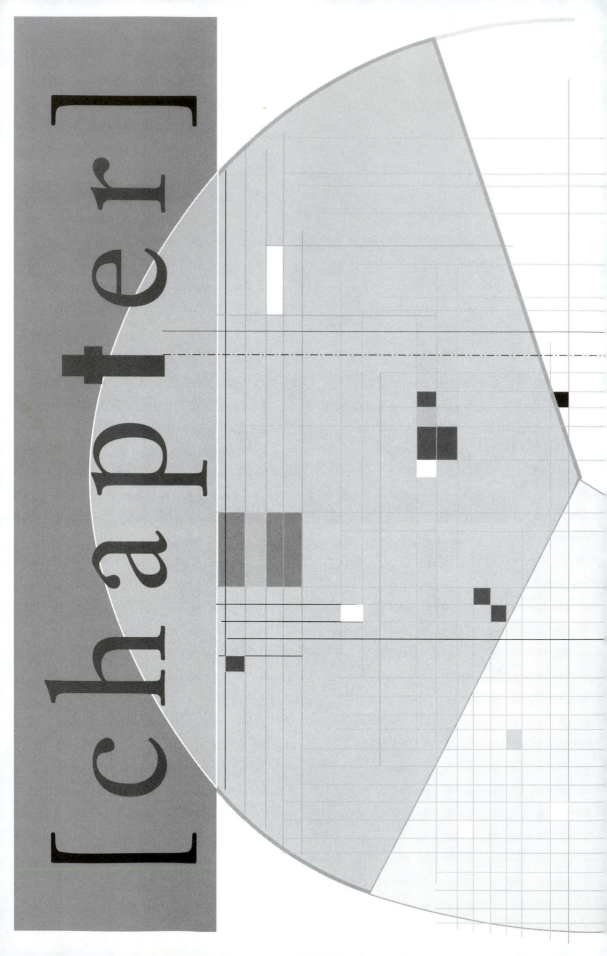

[chapter]

Creating Line-Graph Variants and Scatter Plots

6

Line graphs, layer graphs, and scatter plots all have an L-shaped framework, and recommendations for producing the framework (and labeling and titling it) were given in Chapter 3. Now follow the step-by-step recommendations in this chapter to create the content of these formats.

Line Graphs

Line graphs, because of the continuous nature of their content, are particularly useful in showing how quantities change over time.

● **recommendation** ───
Vary the salience of lines to indicate relative importance.

The most salient line will be noticed first and interpreted as the most important (the principle of salience at work). If you are looking at all three networks' morning shows and comparing their ratings, the graph on the left is the style you want. But if your subject is NBC's "Today" and you want to emphasize that show's ratings against the other two shows in the field, the one on the right better illustrates your focus.

● **recommendation** ───●
Ensure that crossing or nearby lines are discriminable.

When lines are nearby or cross often, special care must be taken to ensure that they are discriminable. One way to increase discriminability is to use dashed lines; another way is to use different colors (see Chapter 7).

● **recommendation** ───
Dashes in lines should differ by at least 2 to 1.

To be immediately discriminable from one another, dashed lines should differ in elements per inch in a ratio of at least 2 to 1. For example, if one line has 4 dashes to the inch, no other line should have more than 2 or fewer than 8 dashes to the inch. This recommendation is based on the principle of *levels of acuity*: The visual system registers information at different levels of scale (see pages 3–5). If the differences in levels are roughly 2 to 1, they will be processed by different "input channels"; variations less than that will be processed by the same channel and the distinctions between them not readily perceived. If you have so many lines that it is not possible to maintain a 2-to-1 ratio of elements per inch, use lines like those illustrated in **do**; it has been shown that these lines are perceptually highly discriminable.

140

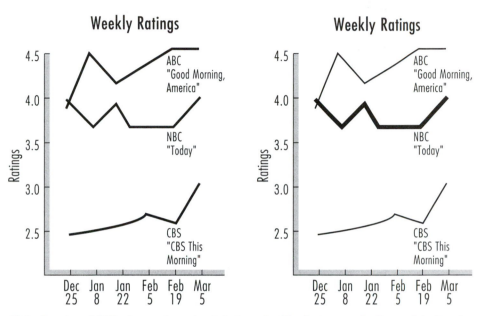

If the focus is on NBC, the graph on the right is preferable; the increased salience of the line for NBC immediately draws the reader's attention.

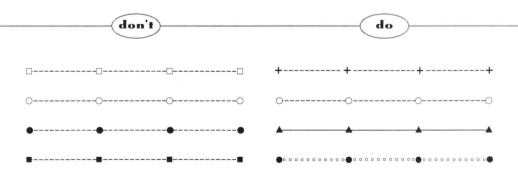

Studies have shown that the four lines and symbols on the right are highly discriminable.

If lines connect discrete points, the points should be at least twice as thick as the line.

Some line graphs connect discrete points that mark specific levels on the X axis. These points should be specified by dots or symbols, which should be at least twice as thick as the line. Again, the principle of levels of acuity is at work: At least 2-to-1 ratios of size (dot-to-line) are desirable when relatively small marks must be distinguished immediately. The **do** version follows this recommendation.

Use discriminable symbols for points connected by different lines.

When lines are nearby or cross often, discriminability can be enhanced by using visually distinct symbols for important points—either in conjunction with different dashed or colored lines or with solid lines. The symbols used must be as discriminable as possible. Plus signs, Xs, open circles, and filled triangles remain distinct even when reduced to small sizes. The difference between filled and unfilled version of the same element, however, often is difficult to discern after photoreduction.

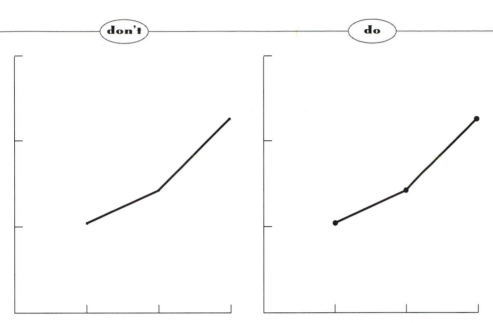

The points over the tick marks are especially important and should be emphasized by discriminable dots.

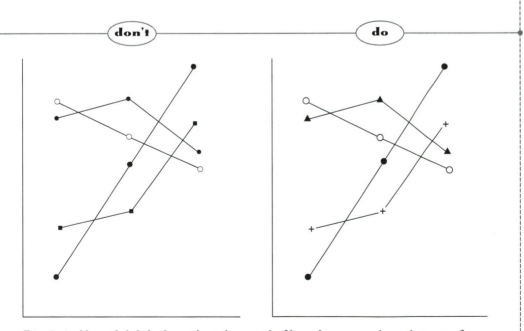

Discriminable symbols help the reader to keep track of lines that are nearby or that cross often.

Do not fill in the area between two lines to emphasize relative trends.

At first thought, it might seem like a good idea to fill in the area between two lines in a line graph, thereby creating a shape, as in **don't**. In accordance with the principle of compatibility, features of the shape itself will convey information about the relationship between the two functions: Shapes that are larger on one side than the other tell an instant story about the relative changes in the two levels of the parameter. Also, the large distinctive area will be salient, drawing the reader's attention to this relation. There are two potential major drawbacks to this practice, however. First, the filled-in area may be salient to such a degree that the reader will ignore other lines in the display. Second, the reader may mistake a line graph for a layer graph and attempt to read a cumulative total, as one would in a layer graph, not the absolute independent values indicated by two or more lines on a line graph.

Ensure that error bars are discriminable by using half-"I" bars or by staggering them.

The error variation around an average should be illustrated with a small "I" bar on the dot that represents the average. The distance above and below the dot indicates the likely range of variation if the data were collected again in the same circumstances (see Appendix 1). If the "I" is centered on the point, it is grouped with it by the principles of proximity and good form. If error bars happen to overlap, include only the top half of the upper bar and the bottom half of the lower bar, as illustrated in **do** (above). This way the bars will be more discriminable, the display less cluttered. If the error bars from different points overlap so much that even truncating them will not render them discriminable, then consider staggering the points slightly, as in **do** (below). Overlap is part of the message, providing information about the ranges of variation, but the reader should be able to tell which error bars belong to which points. Finally, in some cases the error bars are so large that they threaten to overwhelm the trend data; this may be appropriate if the error is so large that the trends are not significant. If the trends are to be taken seriously, however, you may want to make the lines connecting the dots more salient than the lines used to draw the error bars. This recommendation has been followed in all the examples.

144

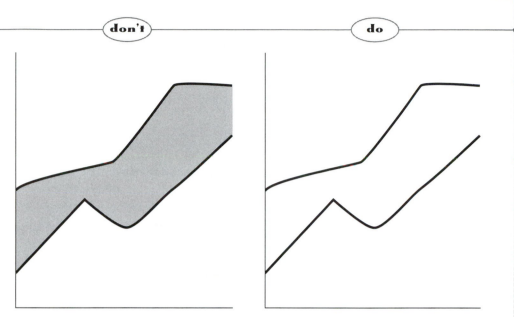

The filled-in region is not only distracting, it also may mislead the reader into thinking this is a layer graph.

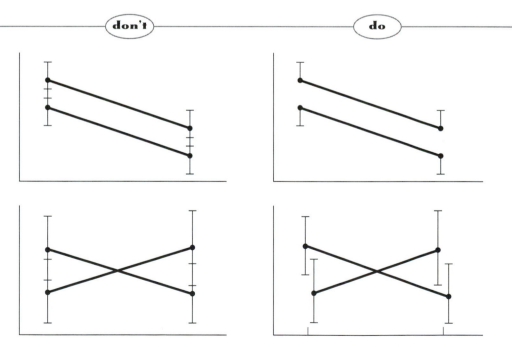

Unless your message is that variability is so large that points are not really distinct (which is unlikely if you want to graph them), use the techniques illustrated at the right to make error bars discriminable.

In mixed line and bar displays, make one function primary.

If two types of data are strongly related, you may want to include them in the same display even if a different format is appropriate for each type. Such two–dependent-measure displays are apt to be complex, and so they are most effective if one component is made secondary to the other, leading the reader to take in the display one part at a time. "Business Registrations and Failures" includes a line to indicate the number of new businesses (one dependent variable) in each of several years (the independent variable), and a set of bars to indicate the number of businesses that went under (another dependent variable) in each of those years. The designer wanted to emphasize that the number of new businesses was steadily climbing and so used a line graph for that variable, while the bar graph component allows the reader to read off point values for the number that failed. Because the editorial emphasis is on the positive aspect of the data, the number of start-ups is visually emphasized. As in this example, the more important dependent measure—in this case, new businesses—should be presented on the left Y axis, and the corresponding data should be plotted near the top; because we read from left to right and top to bottom, information in these positions is most likely to be attended to first. Furthermore, it is useful to make the more important information more salient. The scales for the dependent measures can be adjusted so that one function can be placed above the other, either by choosing appropriate distances for the tick marks or by excising part of the axis (see pages 78–81 and 84–88).

Follow the guidelines for labels given in Chapter 3.

Ensure that labels are easily detectable; avoid fonts in whisc letters share many features; use visually simple fonts; words in the same label should be close together and typographically similar; the more salient labels should label the more general components of the display; use the same size and font for labels of corresponding components; use the same terminology in labels and surrounding text; label each component of the content material; label content elements directly. These recommendations are illustrated and discussed in detail on pages 88–101.

Business Registrations and Failures in England and Wales

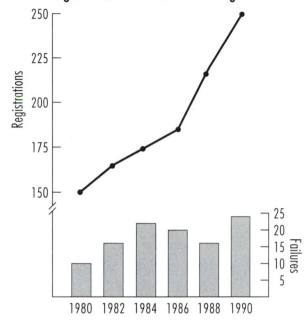

The heavy black line draws the reader's attention to business registrations, the main topic of the display.

Labels of all lines should appear in the same part of the display.

Look at **don't** and decide which group is growing more quickly, Liberals or Others. This is more difficult to determine than it has to be because of the way the labels are placed. Whereas **do** uses the principles of perceptual organization to pair labels and lines, **don't** is a victim of these principles. Labels of different lines should line up, forming a column; otherwise, the viewer will have to search for the different labels in different areas of the graph. The principle of informative changes leads readers not to expect changes unless they carry information. Pick an area where the lines are least cluttered (recall the principle of discriminability) and put all labels in corresponding positions.

Position labels at the ends of lines.

The previous recommendation often implies this one; the only reason not to label lines at their ends is if the labels are too large to fit into the available space. Putting a label at the end of each line is a good practice because the principle of good continuation will serve to group the labels with the lines.

Use hierarchical labeling.

A hierarchical labeling scheme using different levels of labels minimizes the effort required of the reader. Look at **do**: Are sales of both condos and single-family homes increasing more in one community than the other? The hierarchical labeling scheme allows the reader to compare the data along individual dimensions (dwelling type, location) and eliminates re-dundancy and its attendant clutter. If the lines are not arranged to permit hierarchical labeling, either fully label each line directly with the name of the independent variable and level (keeping the order of the two names constant) or—if this produces a cluttered display—use a key. If each infor-mation-bearing element is to be labeled, these are the only choices.

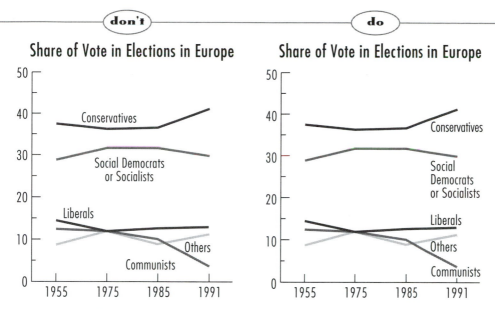

Searching for labels is time-consuming; the arrangement on the right is not only more efficient, it is also less busy visually.

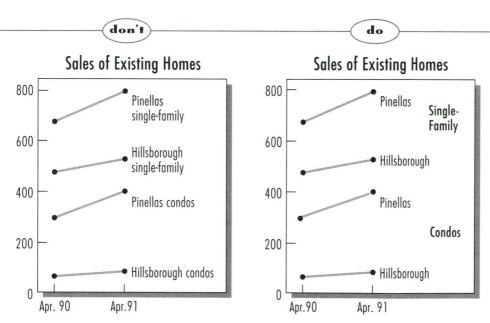

A hierarchical labeling scheme helps the reader to compare data along one dimension (here, dwelling type or location) and reduces clutter.

Label critical point values explicitly.

The primary purpose of line graphs is to illustrate specific trends or interactions among values. If you also want to emphasize a few specific point values, those values can be labeled directly, as in **do**. However, respecting the principle of relevance and limited processing capacity, it is best not to label specific point values unless they are particularly important. For example, if the display is supposed to show the peaks and valleys of real estate prices, and the precise amounts of the absolute high and low are important (rather than simply the visual impression of their relation), put only those two numbers on the display. The numbers should be close to the appropriate parts of the line (so they are grouped in accordance with the principle of proximity) and the title should indicate what the numbers mean.

Layer Graphs

A layer graph presents cumulative totals. For example, this format could be used to show government assistance to depressed areas in the United Kingdom from 1980 through 1990, divided by the four different funding programs. The height of the top line would indicate total funding. This graph would include four layers, formed by the three lines below the line indicating total funding, and each layer would represent the amount of one of the four types of grants. These displays are like stacked-bar graphs in that they show cumulative totals; they are like line graphs in that they show trends. (In a sense, the stacked bar is a snapshot, the layer graph a movie.) Because layer graphs are a hybrid of line graphs and stacked-bar graphs, most of the recommendations offered here follow from recommendations made earlier for these other formats. Any possible confusion between a line graph and a layer graph can be eliminated by proper titles, labels, and captions.

Layer graphs differ from stacked-bar graphs in only two important ways: The X axis must specify an interval scale, and there are no vertical lines demarcating discrete intervals for levels along the X axis. The content elements are lines that define filled-in regions.

don't · do

Median Real Estate Prices

Median Real Estate Prices

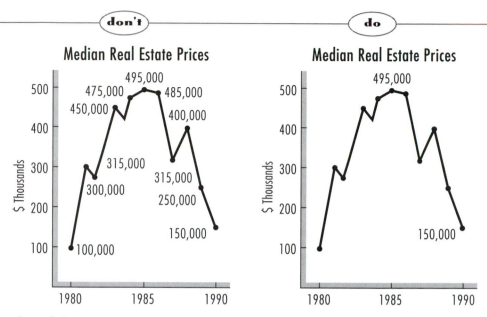

A graph that provides more information than readers need forces them to filter and search.
Directly label only the critical values.

● **recommendation** —————————————————————————————————————

Use shading to ensure that a layer graph is not misread as a line graph.

Without shading, it is not immediately clear whether **don't** is a line graph or a layer graph—whether the lines represent independent trends considered as parallel events (line graph) or boundaries of areas that sum to a total represented by the height of the top line (layer graph). Any doubt is removed in **do**.

● **recommendation** —————————————————————————————————————

Put the layers that change least at the bottom.

By putting the layers that change the least at the bottom, you can make it easier for the reader to compare the degree of change in specific segments. This recommendation is analogous to those illustrated on pages 113 and 129.

● **recommendation** —————————————————————————————————————

Follow the guidelines for labels given in Chapter 3.

Ensure that labels are easily detectable; avoid fonts in which letters share many features; use visually simple fonts; words in the same label should be close together and typographically similar; the more salient labels should label the more general components of the display; use the same size and font for labels of corresponding components; use the same terminology in labels and surrounding text; label each component of the content material; label content elements directly. These recommendations are illustrated and discussed in detail on pages 88–101.

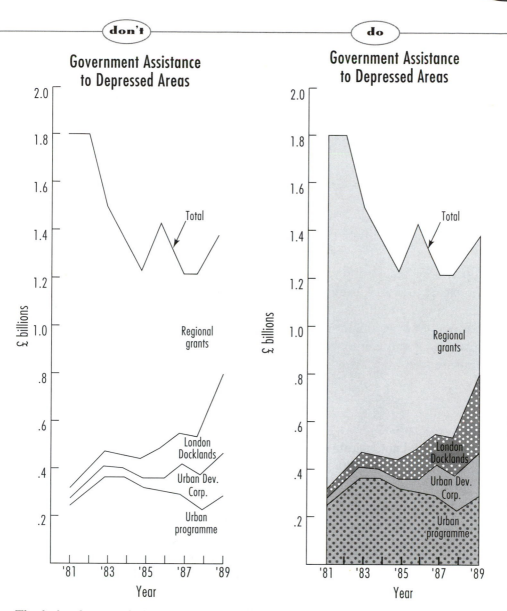

The shading leaves no doubt that this is a layer graph, not a line graph.

Scatter Plots

Scatter plots employ dots and symbols to represent discrete points, each of which typically corresponds to a single measurement. These displays are useful for spotting trends in data. In many respects, scatter plots are like line graphs without the lines, so most of the recommendations that pertain to line graphs also apply here.

• **recommendation** ──•

Ensure that point symbols are discriminable.

If points represent different things, ensure that the symbols for such points are clearly distinct, even if the display is reduced. As in line graphs, +, X, open-circle, and filled-triangle symbols are desirable because they remain distinct even at small sizes. The difference between filled and unfilled symbols of the same shape often is difficult to discern, especially after photoreduction.

• **recommendation** ──

Do not indicate overlapping points with different symbols.

Where points overlap, the **do** graph uses larger dots, whereas the **don't** version uses different symbols. Which is easier to read? As illustrated in **do**, a darker or larger dot, or a double dot, better conveys the impression that there is more of something than does the difference between a triangle and a circle. The physical characteristics of the marks should be analogous to the quantitative information, as dictated by the principle of compatibility. Recall, however, that the visual system distorts differences in area. Use at most four differences in dot size to indicate different ranges of overlapping points; the sizes should differ by multiples of two to be immediately discriminable. A key or caption stating the values of the different symbols is necessary.

don't

do

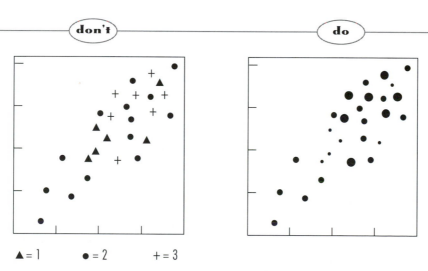

▲ = 1   ● = 2   + = 3

Different-sized dots exploit the principle of compatibility, using more-is-more to indicate overlapping points; the use of different symbols requires the reader to memorize a completely arbitrary convention.

Error bars should not make less stable points more salient.

If it is important for your message to include error bars, then put a light circle around each point or a light "I" error bar, as in **do**, to indicate a range of variation; the error indication must be light, or less stable points (those with larger variability) will be specified by a larger visual change and hence will be more salient—which is exactly what they should not be, because they are in fact less, not more, reliable.

Ensure that best-fitting lines are discriminable and salient.

Best-fitting lines often are supplied in scatter plots to indicate the trend in the data. In the example, the Y axis specifies the number of arrests in 1990 and the X axis the number of arrests in 1991, and each dot represents a convicted criminal. For any good-sized city, you would have a cloud of points. You then might fit a line through the cloud (see Appendix 1), which would rise from left to right. If two levels of a parameter (say, men and women) are plotted, and best-fitting lines are provided for each, make sure that the lines are easily discriminable (but see pages 46–47 before plotting more than one data set). It is not a good idea to use patterns of dots and dashes to distinguish the lines because the patterns may be confused with the point symbols. Different colors are best, if possible.

If you include a best-fitting line or lines, ensure that the summary statement your graph makes is salient by making the line or lines at least as heavy as the points, as in **do**. The major purpose of a best-fitting line is to ease the processing load on the reader, and the line will not serve this end if it is not seen quickly and easily.

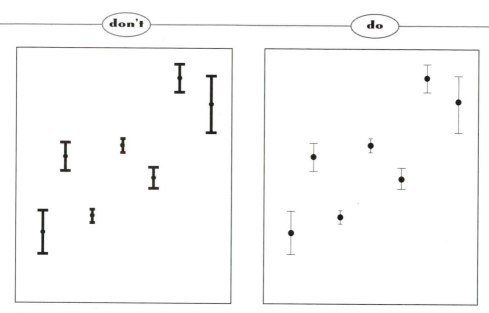

Heavy error bars in fact draw the reader's attention to the least reliable points, those that have the largest bars.

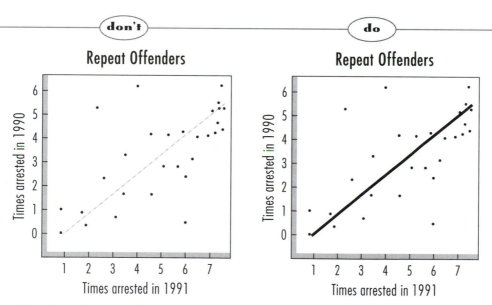

A best-fitting line cannot be a good summary statement if it is not immediately visible.

• recommendation ──────────────────────────────────•

If labels are used, follow the guidelines given in Chapter 3.

Ensure that labels are easily detectable; avoid fonts in which letters share many features; use visually simple fonts; words in the same label should be close together and typographically similar; the more salient labels should label the more general components of the display; use the same size and font for labels of corresponding components; use the same terminology in labels and surrounding text; label each component of the content material; label content elements directly. These recommendations are illustrated and discussed in detail on pages 88–101.

• recommendation ──────────────────────────────────

If more than one best-fitting line is present, label each directly.

Include a label in the display for each best-fitting line, as in **do**. Position the label closer to the line it labels than to any other line, using proximity to produce the appropriate grouping.

• recommendation ──────────────────────────────────•

Identify the method of fitting the line.

If a best-fitting line is included in the display, identify the method of fit (see Appendix 1) in either the title or the caption. The reader will not be able to interpret the meaning of the line without this information.

•–·→

The Next Step

Turn to Chapter 7 for recommendations on shading, hatching, color, three-dimensional effects, inner grids and other optional features.

don't   do

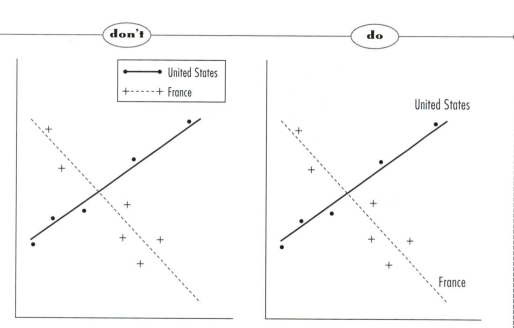

If the data permit you to plot more than one set of data, ensure that best-fitting lines are labeled, preferably directly.

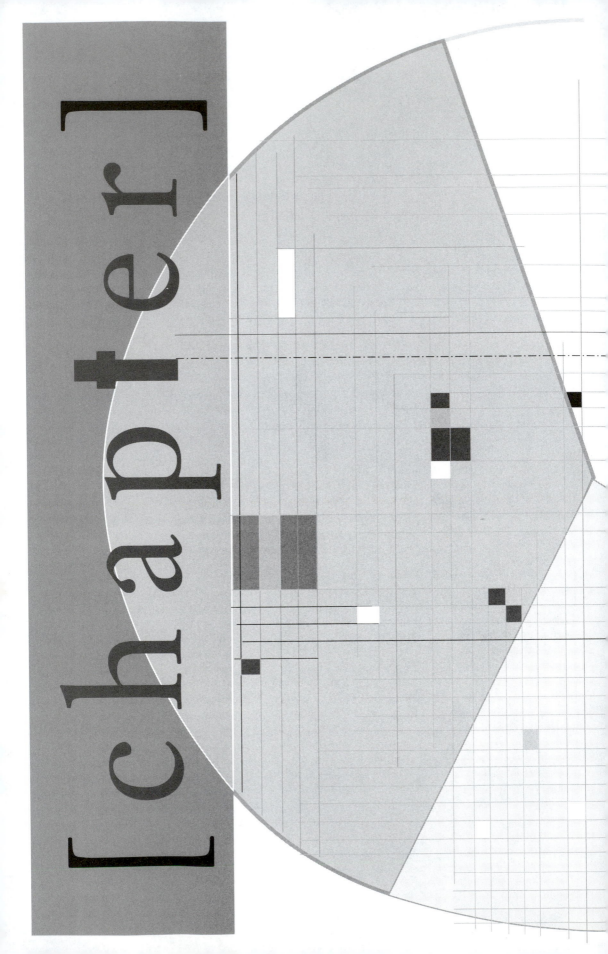

[chapter]

Creating Color, Filling, and Optional Components

7

You have selected a display type and created the display. However, even if you have followed the recommendations, the components may be difficult to discern and therefore the graph difficult to read. Hatching, shading, or color can be used to make regions—wedges of a pie, bars, layers—distinctive. Color, as we will see, has further applicability. These techniques not only make a display easier to read, they also enhance its visual appeal. Three-dimensional effects can also make the display more attractive, but unless used carefully they can muddle the message. Other ways of clarifying your display may be an inner grid, background, caption, or key. Recommendations for creating these features (as well as for multipanel displays, which consist of more than one graph) are given in this chapter.

Color

Color not only adds visual interest to a display but also can help you to communicate effectively. If used improperly, however, variations in color can become an active impediment to clear communication. Color is an enormously complex topic, for a number of reasons. First, color is not a single entity but can be broken down into three aspects. The *hue* (what we usually mean by color) is its qualitative aspect, which depends on the wavelength of the light (from long at the red end of the spectrum to short at the violet end); the *saturation* is the deepness of the color (which can be varied by the amount of white that is added); and the *intensity* is the amount of light that is reflected (if the display is projected from a slide or shown on a TV screen) or emitted—in either case, intensity can be varied by the amount of gray that is added. Second, the perception of hue depends in part on the surrounding colors, and so the display must be looked at as a whole. Finally, people vary greatly in their perception of color. A sizable percentage of the population is color-blind (that is, they have trouble distinguishing certain colors, usually red and green); also, people's assessments of how intense two colors must be to seem equally bright vary greatly. The following recommendations will help you to ensure that the colors you use will be discriminable by everyone, that they do not irritate the reader, and that they convey information properly.

• **recommendation** ─────────────────────────────────────
Use colors that are well separated in the spectrum.

To increase discriminability, colors that are juxtaposed in your display should be ones that are separated by at least one other noticeably distinct color in the spectrum, depicted in the color wheel. Color is registered initially by three types of receptors in the eye (which are most sensitive to the wavelengths that are seen as red, green, and blue). We can perceive a gigantic number of colors because each type of receptor responds to a wide range of wavelengths, and typically all three respond to some extent to any wavelength, but with different degrees of enthusiasm. It is the mixture in outputs from all three kinds of receptors that conveys the precise color. Two colors that are produced by a similar mixture will not be discriminated. The six colors that we perceive as being most separated are reddish-purple, blue, yellowish-gray, yellowish-green, red, and bluish-gray. The "eleven colors that are never confused" are white, gray, black, red, green, yellow, blue, pink, brown, orange, and purple. Nevertheless my strong advice is not to use all eleven in a single display: Aesthetics aside, people can generally keep in mind the distinctions among only nine colors.

The color wheel, which represents our psychological perception that color is not a continuum.

Adjacent colors should have different brightnesses.

Our visual systems have difficulty registering a boundary that is defined by two colors that are of the same brightness, as in **don't**. Brightness is the psychological impression of intensity, but whereas intensity can be measured by a light meter, our eyes and minds do not accurately translate intensity into brightness. When colors have the same objective intensity, we see blue as the brightest color, followed by red, green, yellow, and white. Brightness must be adjusted subjectively, and it should vary by quite a bit, for two reasons. First, people vary considerably in their perception of how bright a color is. Second, different sorts of room lighting affect the brightness of different colors in different ways.

Make the most important content element the most salient.

According to the principle of salience, larger differences will be noticed first, so ensure that the most important wedge, object, or segment stands out the most. If no one element is most important, all should be equally salient. After you adjust the brightnesses so that the colors will be easily distinguished, adjust the saturations for the different hues you are using until none predominates.

Use warm colors to define a foreground.

Any visual stimulus projects slightly different images into each eye because the eyes are in slightly different places. By the process of stereo vision, the disparity between these images is used by the brain to infer distance. Because each eye is actually aimed slightly askew when we look straight ahead, the lens of the eye acts like a prism. Incoming light is thus broken into the colors of the spectrum, which are projected to slightly different locations on the retina. The psychological effect is similar to that sensed when one object is actually closer than the other: Some colors are perceived as "closer" than others. A "warm" color, such as red or orange, will appear to be in front of a cooler one, such as green, violet, or black. This effect can be advantageous if used in accordance with the principle of salience to highlight a feature of the display. If misused, however, the effect can be distracting: A warm-colored object drawn behind a cool-colored one will struggle to move to the foreground, as in **don't**.

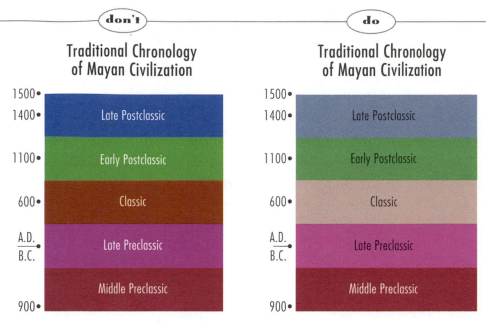

don't do

Equal brightnesses make boundaries harder to discern. A given level of intensity appears of different brightnesses to different people, and so differences in intensities must be large enough that all viewers will perceive distinct differences in brightness.

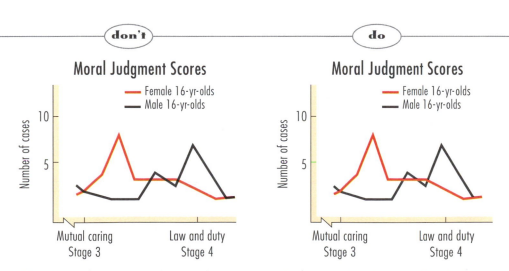

The red line that appears to be struggling to move to the foreground produces an effect that is neither esthetically nor functionally desirable.

● **recommendation** ───
Avoid using red and blue in adjacent regions.

The lens of the eye, unlike lenses in good cameras, cannot properly focus two very different wavelengths at the same time. This is why red (a relatively long wavelength) and blue (a relatively short wavelength) will seem to shimmer if juxtaposed, as in **don't**. For most purposes, this effect will distract the reader and should be avoided. Also, it is generally a good idea to avoid deep, heavily saturated blue; the eye cannot focus the image properly (it will fall slightly in front of the retina), and so deep blues will appear blurred around the edges. Similarly, avoid cobalt blue, which is in fact a mixture of blue and red; for people with normal vision, this color can never be fully in focus because the eye cannot accommodate to both frequencies at the same time. The halo you have probably noticed around blue street lights at night is not fog; your eyes are failing to focus the image properly. Finally, it is often a good idea to avoid using red and green to define a boundary because about 8% of the male population have trouble distinguishing these colors.

● **recommendation** ───
Respect compatibility and conventions of color.

Objects sometimes have characteristic colors; do not use color in a way, as in **don't**, that contradicts compatibility. In addition, different cultures have specific conventional symbols, some of which involve color. In the United States red symbolizes stop or danger; blue, coolness, cleanliness, or safety; green, life. Colors may also have political connotations. Know your audience.

● **recommendation** ───
Use color to group elements.

The principle of similarity implies that regions of the same color will be seen as a group. This principle, used in **do**, is very useful if the reader is to compare two or more elements in different places. If two pie graphs are used, using the same color for corresponding wedges will group them effectively. Note, however, that good black-and-white patterns can be almost as good as color distinctions for grouping and identifying individual elements.

The eye cannot properly focus upon red and blue at the same time, so the boundary between these colors shimmers.

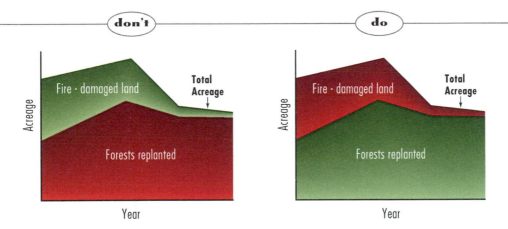

The principles of compatibility and cultural convention extend to the use of color.

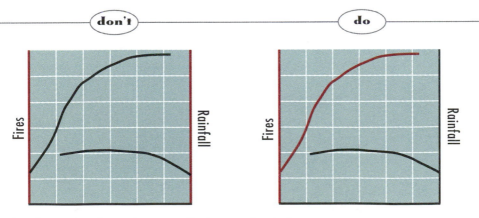

If used properly, color can be a very effective grouping device.

Avoid using hue to represent quantitative information.

Respecting the principle of compatibility, do not use hue to represent differences in amounts; use it as a label to promote discrimination and grouping. Compare **don't** and **do**: Shifting from red to violet does not convey the impression that an amount has been added the way that shifting from a short bar to a tall bar does. In fact, the red-to-violet shift *does* involve more of something (more oscillations in the amplitude of a light wave). But although the wavelengths of the hues order them into red, orange, yellow, green, blue, and violet, hues are not a psychological continuum.

Use deeper saturations and greater intensities for hues that indicate greater amounts.

If practical considerations force you to use hue to convey quantitative information, then use deeper saturations (more color) and greater intensities (more light) for hues that indicate greater amounts, as in **do**. We see increases in both of these visual dimensions as increases in the psychological impression, and so these increases can signal increasing amounts effectively. This recommendation will also lead you to avoid using hue (the qualitative aspect of color), saturation (the deepness), and intensity (the amount of light) to convey different types of information. Do not be tempted to specify three independent variables in a single pie graph (the **don't** version of the ice cream example attempts to use intensity to indicate a third variable, the average temperature). The reader will have a very difficult time sorting them out because hue and intensity, and hue and saturation, are integral perceptual dimensions. A reader looking at a colored wedge cannot help but pay attention to its deepness and intensity as well as its hue. Like height and width, these properties should not be varied independently to convey information about two variables.

Avoid using blue if the display is to be photocopied.

Many of the most popular black-and-white copying technologies reproduce the color blue poorly or not at all. If blue is used, test to be sure that it will reproduce well.

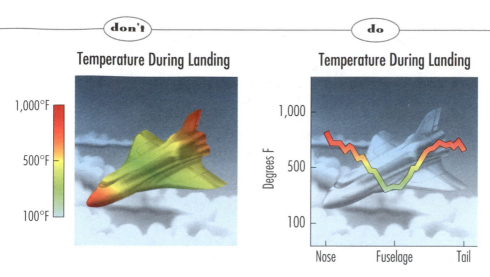

Because differences in hue are not immediately perceived as differences in amount, the reader is required to memorize a key. However, differences in hue can be used to add visual appeal, provided that the principle of compatibility is respected.

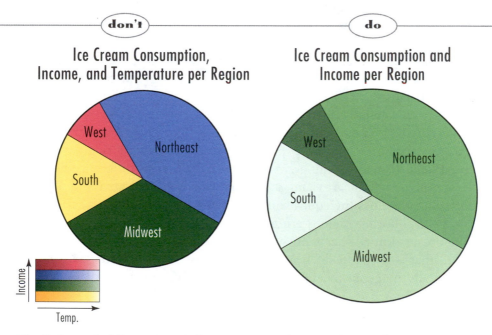

The display on the left uses size to indicate consumption by region, hue to indicate income, and saturation to indicate temperature; it is a puzzle to be solved. The display on the right varies only size for region and saturation for income level and readers can clearly sense the orderings here.

Hatching and Shading

Individual wedges of pie graphs, elements of visual tables, or segments of divided bar graphs are often filled with hatching or shading. We must take care to consider the spacing between texture elements, to ensure both that different regions can be readily distinguished and that the patterns are not visually irritating.

• recommendation ───

Visual properties must be discriminable.

The principle of discriminability states that two or more marks must differ by a minimal proportion in order to be distinguished from one another; the critical factor is the proportion, not the absolute amount (see page 90). When you use intensity (which we see as brightness) or texture (such as cross-hatching) to define different wedges or segments, remember that the visual system does not simply record what it sees point for point, like a television camera. Rather, it tries to delineate edges and boundaries, both between objects and their backgrounds and between individual parts of objects. The brain does this by responding to differences in contrast, color, texture, orientation, and other properties between neighboring parts of the visual field. Whenever the brain registers a significant difference between, for example, a light area and a dark area or a striped area and a polka-dotted area, it does the equivalent of drawing a line that serves as a boundary. If neighboring regions are too close in lightness, texture, or color (as they are in **don't**), the brain will not readily establish a border between them: that is why a polar bear lying on the snow is hard to see. Camouflage is advantageous to a polar bear but not to a display.

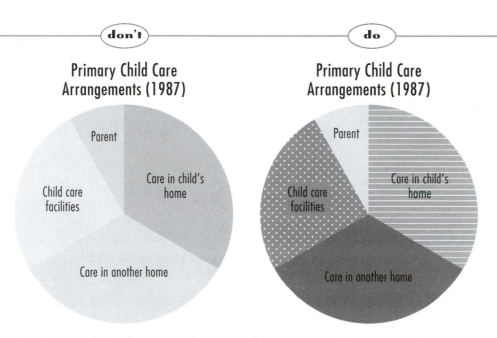

don't

Primary Child Care Arrangements (1987)

Parent

Child care facilities

Care in child's home

Care in another home

do

Primary Child Care Arrangements (1987)

Parent

Child care facilities

Care in child's home

Care in another home

Poor discriminability of regions, confusing enough here, is particularly damaging if the reader is to compare corresponding wedges of a multipanel display.

Orientation should vary by at least 30 degrees of arc.

When you read an analog clock (one with hands, not just numbers), some positions of the hands require closer attention than others. It may be no accident that clock faces are divided into 12 equal and easily discriminable increments. The principle of *orientation sensitivity* states that if differently oriented hatchings are used to distinguish regions, they should be at least 30 degrees apart (the angle formed by the hands of a clock when they point to adjacent numbers, as at 12:05). Various studies of discrimination of lines at different orientations have shown that when lines differ by at least 30 degrees, as in **do**, we can distinguish among them without having to pay close attention. Neurons at various levels of the visual system are tuned to respond only to edges or shapes at particular orientations; in the initial stages of processing, the neurons have rather broad "tuning curves," and so respond to edges within a relatively wide range of orientations (the mind is not a camera). Thirty degrees is large enough to ensure that different input neurons will be used to encode patterns of different orientations.

Vary the spacing of texture patterns with similar orientations by at least a ratio of 2 to 1.

The example illustrates the salubrious effects of alcohol (by some measures) on Londoners. It is immediately obvious from **do** that there are more nondrinkers in poor health than in good health, and vice versa for moderate—and even heavy—drinkers. The **don't** version is not so easy to read. If cross-hatching, stripes, dots, dashes, or other regular patterns have similar orientations (that is, within 30 degrees of each other), the densities of the pattern should differ by at least 2 to 1. If a region has eight hatch lines to the inch, to be immediately discriminable adjacent regions should have either four or fewer lines to the inch, or sixteen or more lines to the inch.

This recommendation is based on the discovery that the visual system operates at multiple levels of acuity (see pages 3–5). What we see comes to us from a number of "input channels," distinct in regard to the size of the region they register (and thus distinct in the amount of detail that they pick up—less detail is detected when larger regions are monitored). Furthermore, there are several types of each channel, tuned for different orientations. All these channels operate at once.

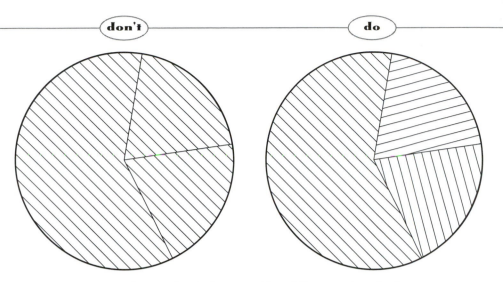

Line orientations must be immediately discriminable to delineate regions clearly.

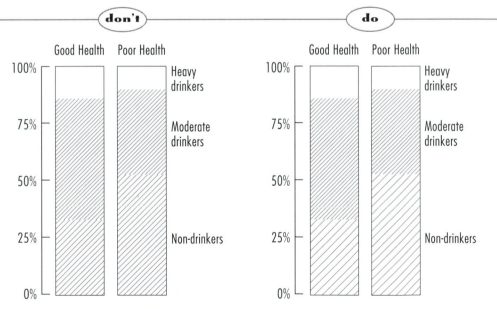

If segments are not immediately discriminable, the reader has to work harder than necessary to compare corresponding elements.

The acuity of a channel is described by its spatial frequency, the number of regular light/dark changes (for example, light and dark stripes of equal width) that fit into a specific amount of your visual field. Spatial frequency is measured in cycles per degree of visual angle. One degree of visual angle is roughly equivalent to the apparent width of your little finger when you hold it at arm's length and look at it with one eye. A cycle is a complete sequence of the light/dark change. A spatial frequency of two cycles per degree, then, would have two dark-light pairs for every extent that is as wide as your little finger appears when held at arm's length.

Each channel, which is tuned to a particular orientation, responds to a range of spatial frequencies of roughly 4 to 1 (from minimum to maximum). Thus, when we look at an area of evenly spaced hatch lines, dots, or patches, we will also mentally take in, like it or not, areas of other similarly oriented lines, dots, or patches whose cycles range from half as frequent to twice as frequent. The closer the spacing is to that of the area we are paying attention to, the harder it is to ignore. Therefore, to be immediately discriminable two patterns that have similar orientations should be registered by different channels, which means they should differ by more than 2 to 1.

• recommendation ───────────────────────────────

Avoid visual beats.

Similarly oriented patterns that fall in the same channel—those whose spatial frequencies do *not* differ by at least 2 to 1—may appear to shimmer, a distracting and irritating effect. This shimmering is analogous to what happens if two notes are only slightly different, so that their harmonics will periodically reinforce each other, causing pulsating beats. This is a helpful cue if one is trying to tune a guitar but an unwelcome irritation if one is listening to a song being played. Visual shimmer, illustrated here, results when brain cells try to organize input into homogeneous regions, an attempt that is not clearly successful or unsuccessful if the patterns are very similar but not identical. There is no simple way always to predict when juxtaposed hatch marks will appear to oscillate, creating an op art pattern. If you ensure that the spatial frequency spread respects the principle of levels of acuity, the occurrence of this phenomenon is less likely.

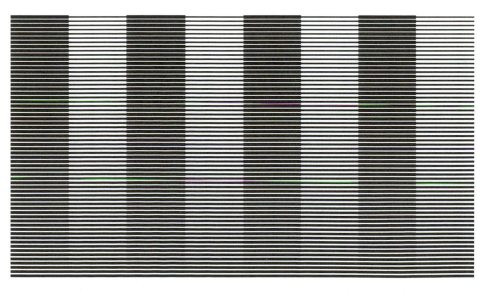

An annoying shimmer occurs when your visual system is struggling to detect a poorly defined edge.

Three-Dimensional Effects

When we look around us, the cues the brain uses to assign depth allow us to see the world as it is—in three dimensions. When the brain, relying on the same cues, addresses a two-dimensional surface, it can be fooled into seeing a third dimension that is not present. Three-dimensional displays, in which depth is indicated along an unseen Z axis, can be useful if they bring the reader's attention to the important comparisons being made, but they can be harmful if they obscure the message. You must decide whether the added visual interest is worth the risk that information will be lost.

A sheet of paper, or a screen, is two-dimensional, and so special tricks must be used to fool the eye and mind into seeing a three-dimensional object. Three types of tricks are often used in visual displays, all of them based on environmental cues that usually signal information about depth. One such cue is linear perspective. The brain interprets converging lines on a two-dimensional surface as if they were parallel lines extending in depth. The second cue is a texture gradient (a progressive change in the density of a texture). In a photograph of a field of wheat or a brick road receding into the distance, the texture—the density of the stalks or bricks—increases in the portions of the scene that are in reality (though not in the photograph) farther away. Our brains assume that as the texture elements become more tightly packed together, the scene is more distant. The third cue is occlusion. Closer objects can obstruct our view of farther objects, but not vice versa—and so our brains use partial occlusion to determine relative distance. If a completely closed shape interrupts the boundary of a second, then the first shape is seen as standing in front of the second, creating a sense of depth.

You can often exploit these cues to convey depth effectively on two-dimensional pieces of paper or computer screens. The principle of *three-dimensional projection* states that patterns will be seen as three dimensional whenever possible. Specifically, the brain operates as if it unconsciously assumes that shapes have regular, closed contours composed of parallel and perpendicular sides, that objects have the same shading and markings all along their surface, and that the world is made of cohesive surfaces rather than clouds composed of tiny drops or shards.

Moreover, the brain operates as if it assumes that the same-sized image (size measured in degrees of visual angle) produced by something farther away signals a larger object. Thus, in the illustration the bar on the right looks larger than it would if it were not partially occluded because one unconsciously compensates for perspective effects.

These cues on two-dimensional surfaces are imperfect substitutes for the real thing, and confusion may lie ahead. The famous drawings of M. C. Escher show how we can be tripped up. Although three-dimensional graphs are rarely as confusing as Escher drawings, they sometimes lean in that direction.

176

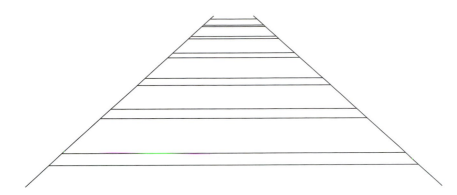

Our brains interpret converging lines as parallel lines extending in depth.

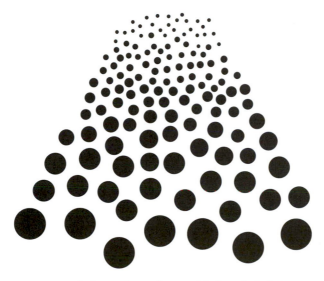

Our brains interpret progressively smaller and more tightly packed elements as constant-sized elements extending in depth.

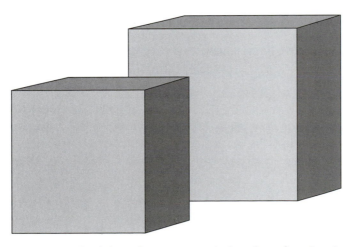

Our brains interpret an enclosed shape that interrupts the boundary of another shape as being in front.

Use views that allow the reader to see the entire content.

Occlusion, one of the depth cues, can result in loss of information. Make sure that all critical aspects of the content (such as the tops of bars) are visible. Avoid occlusion if it distorts the visual impression of the size of a region. It often is best to depict the display as if it were seen from the front, from a position elevated just enough to see over the tops of the "nearer" elements. Avoid adopting a point of view that is too high; top views produce foreshortening, which obscures the relative heights of content elements. And if the point of view is too low, the closer elements obscure the farther ones. It often is useful to expand the height of the graph (which is the same as contracting the X and Z axes); the graph will appear more three-dimensional, and differences in height will be easier to discriminate.

Show all parts of the display from the same viewpoint.

Can you determine from the **don't** graph the average salary of major-league baseball players in 1985? One of the reasons this is difficult is that the content and background are seen from different points of view. Depict all parts of a three-dimensional display from the same viewpoint; the visual system assumes that an object is seen from only a single vantage point, and considerable effort will be required to reorganize the display mentally if it is not. Similarly, never have one part of the display drawn in two-dimensions with a three-dimensional part tacked on. If these recommendations are violated, it will be very difficult to read a value off the display.

Do not use three-dimensional perspective to communicate precise information.

Although three-dimensional graphs can convey general impressions of trends, they are not very good for representing precise amounts. Because two-dimensional displays do not exploit all of the depth cues we use in everyday perception (such as relative motion, stereo vision, shading, and so forth), they do not depict three-dimensional information very accurately. The impoverished depth cues in drawings make it difficult to estimate the depth of the content material and then to extrapolate to the corresponding part of the Y axis. Also, the visual system does not estimate volume very accurately, so if your display is supposed to convey precise information, keep it two-dimensional.

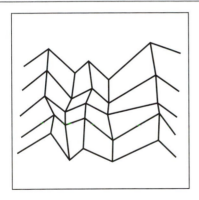 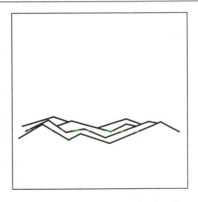

Three-dimensional surfaces should not be depicted from an angle so high that depth is lost (left), or from an angle so low that closer surfaces obscure farther ones (right).

don't   do

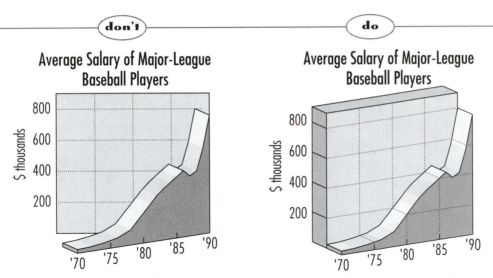

Average Salary of Major-League Baseball Players

Average Salary of Major-League Baseball Players

Values can be nearly impossible to read when all components of the display are not viewed from a single perspective.

● **recommendation** ──────────────────────────────────────

Avoid see-through displays.

See-through displays like **don't** show what we would see if we had X-ray vision; nearer surfaces are treated as if they were transparent, allowing farther surfaces to be seen through the nearer ones. These displays are rarely effective because the reader organizes the parts incorrectly; the principles of perceptual organization result in spurious groupings. Use a different color, shading, or line thickness for the underside of a surface if any part of it is visible. Make the walls of a surface opaque, as in **do**, by shading or by coloring them in a solid color.

● **recommendation** ──────────────────────────────────────

Use a vertical wall for the Y axis.

It is almost impossible to read off the number of major corporate headquarters from **don't**, whereas it is possible (with work!) to do so from **do**. Use a vertical wall for the Y axis, which will appear as a plane projecting in depth along the Z axis, instead of a single line. But even in this case, the reader will be able to read specific Y values only with considerable effort; the three-dimensional cues are not good enough to allow the reader to locate the appropriate position in depth easily. If the reader is supposed to extract precise values, label these values with numbers.

● **recommendation** ──────────────────────────────────────

If the content is a three-dimensional sheet, cover it with regularly spaced grid lines.

Some three-dimensional graphs include a different independent variable along the X axis (running perpendicular to the line of sight, as usual) and along the Z axis (running in depth). The content is a sheet, which typically appears as a landscape with peaks and valleys. Because the local convergence of lines is a good depth cue, the three-dimensional effect will be enhanced if the sheet looks like a surface that, as in **do,** is covered with a relatively dense grid pattern.

don't do

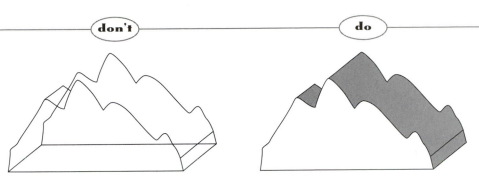

See-through displays often lead readers to group components incorrectly and are generally difficult to interpret.

don't   do

Major Corporate Headquarters

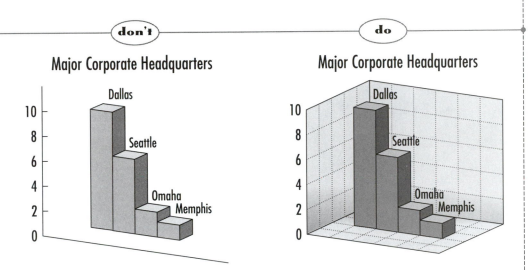

A vertical wall allows readers to extract more precise values than would be possible without it.

don't   do

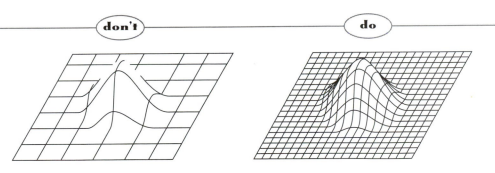

Grid lines can create a three-dimensional surface, but the effect falters if the lines are not packed closely enough.

Inner Grid Lines

An inner grid can be used with any format that includes a framework. Vertical grid lines help readers to isolate a specific place on a line, and horizontal lines help readers to spot specific values of the dependent measure.

• **recommendation** ────────────────────────────────────

Inner grid lines should be relatively thin and light.

Which version of "Champagne Sales" would you rather use to find out how much bubbly was sold in 1986? The inner grid lines should be discriminable from those that outline the framework or content elements. They should not be so salient that they distract, nor should they obscure the content material.

• **recommendation** ────────────────────────────────────

Use more tightly spaced grid lines when greater precision is required.

The principle of relevance implies that grid lines should specify only the level of precision necessary to answer the appropriate questions. As a rule of thumb, consider locating the lines at increments that are about four times the acceptable rounding error, as in **do**. If you want the reader to be able to extract figures to the nearest week, grid lines placed at monthly increments, as in **do**, should be sufficient. We can estimate length quite well (see notes keyed to pages 24 and 28) and can visually divide a line into four roughly equal segments.

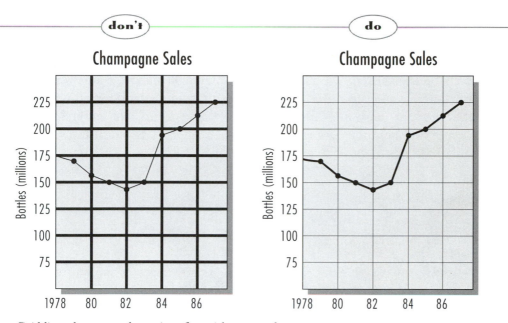

Grid lines that are too heavy interfere with content elements.

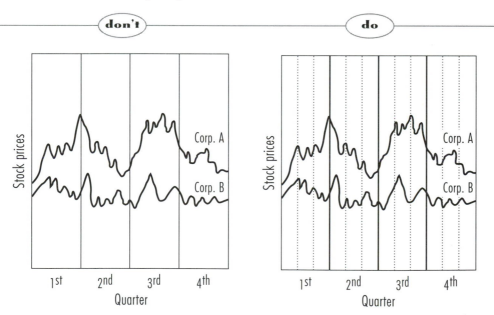

The frequency of the grid lines should be consistent with the fluctuations of the curves; if the point is to present quarterly, not weekly, prices, the data should have been averaged and a single number plotted for each corporation for each quarter.

Insert heavier grid lines at equal intervals.

Which version of "Children in Poverty" is easier to use to determine the years in which 20% of children (up to the age of 18) were under the poverty line? If tick marks vary in size or darkness, you should line up a relatively heavy grid line with each darker tick mark. If grid lines are closer than 1 cm apart, it is a good idea to make some of them heavier. Given our convention of using Arabic numerals, which are based on powers of ten, every tenth grid line should be slightly thicker than the others. If greater precision is required, use an intermediate level of thickness for every fifth grid line. Keeping in mind the principles of salience and discriminability, take care that these lines do not obscure the content. Varying line weight this way will help the viewer to distinguish a single line, which can then be used to track from the content material to the Y axis.

Inner grid lines should pass behind the lines or bars.

Grid lines should not obscure the content material. The principle of good continuation ensures that each grid line will appear to progress across the entire display even though it is physically interrupted by the content material, as in **do**.

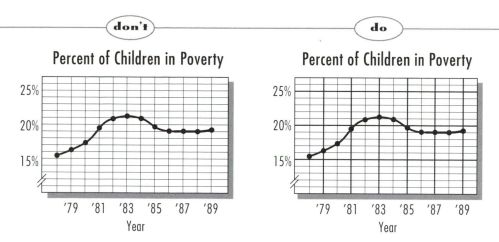

The staggered heavier grid lines help the reader to locate specific values.

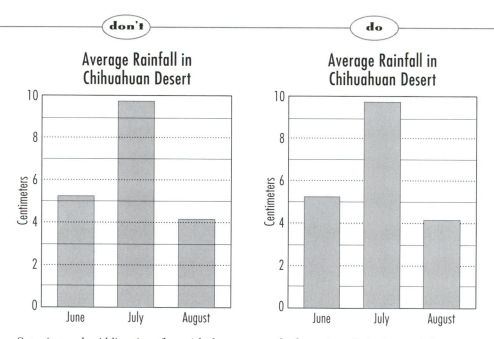

Superimposed grid lines interfere with the content and relegate it to the background; in contrast, we see grid lines interrupted by bars as behind them, and the lines push the bars to the foreground.

Background Elements

The background is not an essential part of a display; it is included primarily to make the display more attractive or interesting. If you choose to use a background, you must be careful to ensure that it does not obscure the message.

• **recommendation** ────────────────────────────────────
Use a background to reinforce the main point of the display.

When a display is used to catch the reader's attention or to enhance a particularly dramatic point, an effective background underlines the message of the display, providing a kind of pictorial label. The background design must be compatible with the content of the display, as in **do**. Show a couple of people the background image you are considering using and ask them to name it with the first word that comes to mind; this word should be appropriate for the message conveyed by the display, both in its direct meaning and in its indirect connotations (recall the principle of relevance). Do not use decorations that convey no information about the topic of the display; according to the principle of informative changes, every element of a display should convey information. Tacked-on elements will only serve to distract the reader.

• **recommendation** ────────────────────────────────────
The background should not be salient.

Can you easily tell from **don't** what houses cost in New Hampshire in 1990? Do you initially have trouble distinguishing parts of the drawing of the house from parts of the content itself? The background should not interfere with the information-bearing lines and regions, nor should its salience lure the eye away from these elements. A background should be lighter and in less saturated colors than those used in the rest of the display, have few details, and have soft edges.

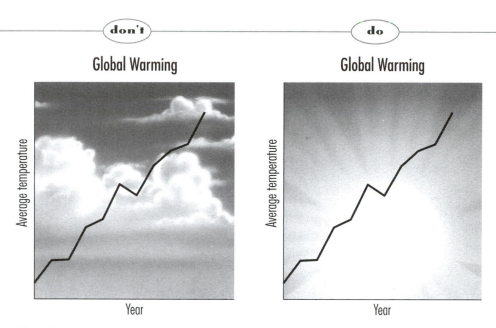

The editorial content of a background should not conflict with the message of the display.

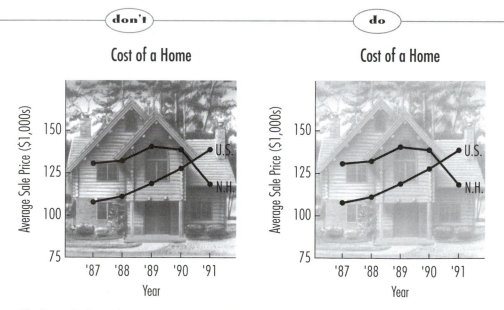

If salience leads readers to see the house before the data, they will have to work to sort out the content.

Background elements must not group with content elements.

Look at the **don't** version: At first glance, it seems that South Korea and India have similar levels of defense spending. But this is not the case, you can see from **do**. Make sure that background figures do not group with parts of the display itself. You should be aware of similarities, proximities, good continuations, common fate, and good-form effects that can produce confusion between background figures and content.

Captions

A caption is not necessary in all contexts. It often is used when graphs are presented in texts, rarely when graphs are used during presentations. When a caption is not included, however, its functions—clarifying and directing attention—must be fulfilled by other material that accompanies the display.

Put the caption under the display.

Convention dictates that captions appear under displays. This recommendation can be ignored, however, if you have many displays and adopt a consistent alternative format (such as placing the caption under the title or in a margin). But if you decide to use an unconventional design, keep in mind that readers will require more time to read the display until they master your scheme.

The caption should be distinct from the text.

The caption should be discriminated from the surrounding text. It should be physically closer to the display than to the text, so that it will group appropriately by proximity. It is also helpful if the caption is typographically distinct from the text, perhaps in a smaller font.

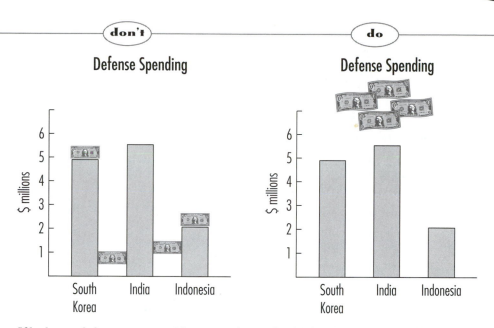

If background elements group with content, the graph is likely to be misinterpreted by a casual reader.

Keys

Use a key to prevent clutter or if direct labeling is impossible (see the discussion on page 56). The key has two components: patches that correspond to the individual components of the content, and labels that identify the patches. The labels should be created in accordance with the recommendations given in Chapter 3 (illustrated and discussed on pages 88–101): Label each component of the content material; ensure that labels are easily detectable; avoid fonts in which letters share many features; use visually simple fonts; words in the same label should be close together and typographically similar; use the same size and font for labels of corresponding components; use the same terminology in labels and surrounding text.

● **recommendation** ──●

Place the key at the top right of a single panel or centered over multiple panels.

In a single-panel display, the key is by convention positioned at the top right; in a multipanel display, it is centered at the top, directly beneath the title. Readers familiar with graphs will expect to find the key in one of these locations. If aesthetic or other considerations lead you to put the key in an unconventional location, realize that the readers may have to search for it, which will tax their limited processing capacity. However, an unconventional location may be acceptable if space is available only in another part of the display, and the key is clearly isolated from the content. In addition, if you consistently place the key in the same location in a number of displays, the reader will soon learn where to find it.

● **recommendation** ──

Labels and patches should be detectable, and corresponding labels and patches should form perceptual groups.

Ensure that the label and the patch—a sample of the texture, the color, or other visual property that identifies the content to be identified—are detectable even after any photoreduction planned. The principles that ensure that lines and regions are discriminable should be respected (pages 162–165 and 170–174). Associate the label and the patch by making sure that they are closer to each other than to any other part of the key; the principle of proximity will then group the elements properly.

don't

do

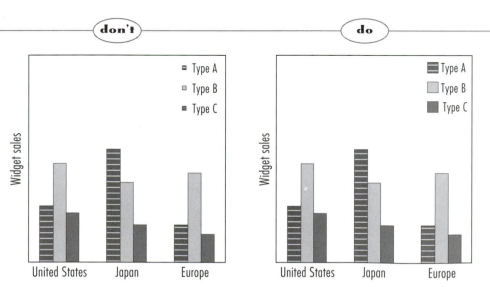

An indecipherable key renders a graph useless.

Use the same order for patches and their corresponding content elements.

The patches in the key should be presented in the same order as the corresponding content elements in the graph itself (in accordance with the principles of informative changes and similarity). Sets of five pictures of milk bottles could be used to illustrate the sales of different types of milk (skim, butter, and so on) in different countries, larger bottles indicating greater sales. The five bottles would be ordered the same way for each country and might be produced with different hatching. The key should then have the patches appear in the same order as the bottles in the graph, presenting the key elements in a column or a row; if a column is used, the top element should refer to the leftmost content element in the display, the second from the top element should refer to the content element second from the left, and so forth. These guidelines are followed in **do**.

Use the content of one divided bar in multipanel displays to serve as the key.

If the components of divided bars are ordered the same way, and are colored or shaded the same way, labels at the left of the components of the leftmost bar may be sufficient to identify the corresponding components of the other panels, as in **do**. This practice is particularly effective if the segments are approximately the same height, so that the horizontal labels will lead the eye across the corresponding segments of the different bars, grouping with them via good continuation, and the bars are close enough to group—and so it is clear that the labels do not apply solely to the leftmost bar. Say you are graphing the contribution of four ethnic groups to the total Democratic vote in each of three states, California, Kansas, and Massachusetts. You would have a separate bar for each state, with the component segments indicating the relative contribution of each group. The bar on the left might be for California, and the four segments would be labeled directly in words at the left of each segment. The components of the two other bars to the right then need not be labeled. This technique reduces clutter (and demands on processing resources) by having the components of the first bar serve double duty as the elements of a key.

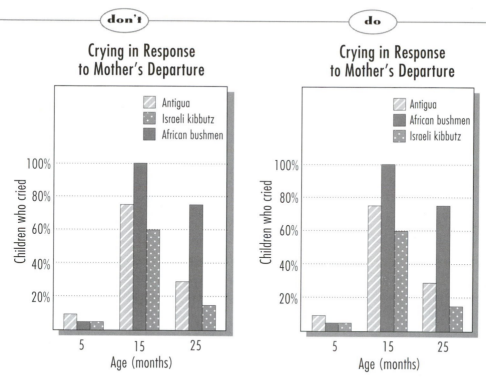

don't
do

Readers will have to search for corresponding bars if the order of the key does not match the order of the content.

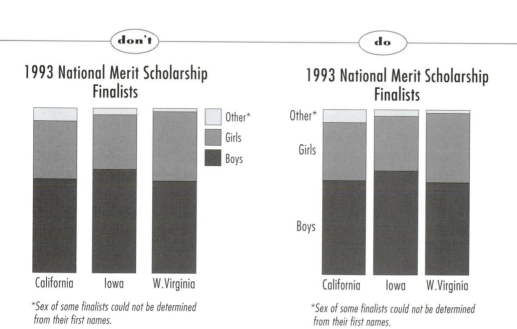

The first of several corresponding bars and its labels can serve as a key for the entire display—if the bars are close enough to group.

Multipanel Displays

Multiple pie graphs or multiple divided-bar graphs can be used to illustrate different levels of two independent variables. One pie or divided bar might correspond to the division of the work force into different jobs in 1980, and another to the division of the work force into different jobs in 1990. Or, you may want to display more than one graph of the same format or a mixture of different types of graphs. The individual displays or panels should be produced in accordance with the recommendations for that type of graph. To create a display with two or more graphs, two major issues must be resolved: Which data should be presented in which panel? How should the panels be arranged in the display?

• recommendation ───

Assign data that answer different questions to different panels.

Data should be grouped so that the viewer can see the important trends and interactions, as is illustrated in **do**. First formulate a concise statement (write it down or say it aloud) of which comparisons are particularly important to convey. Ideally, the reader should be helped to make important comparisons by having those data plotted together. If all comparisons are important, consider the following recommendations.

don't

Computer Simulation of Batting Averages

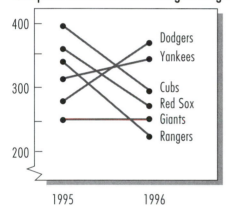

Dodgers
Yankees

Cubs
Red Sox
Giants
Rangers

Computer Simulation of Batting Averages

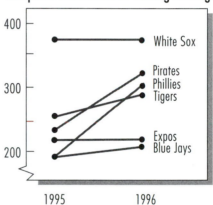

White Sox

Pirates
Phillies
Tigers

Expos
Blue Jays

do

Computer Simulation of Batting Averages
(American League)

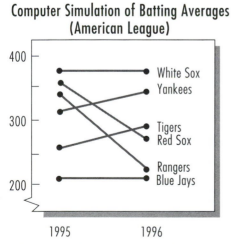

White Sox
Yankees

Tigers
Red Sox

Rangers
Blue Jays

Computer Simulation of Batting Averages
(National League)

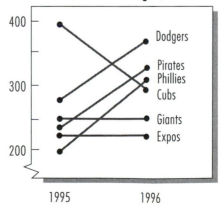

Dodgers

Pirates
Phillies
Cubs

Giants
Expos

Data should be organized to help readers make specific comparisons.

——————————————————————————————————

Assign lines that form a meaningful pattern to the same panel.

Much of the power of line graphs arises from the fact that readers learn the meanings of specific patterns of lines. Unless the point being made requires the reader to compare specific pairs of lines—if so, those lines should be graphed together—assign lines that form a common pattern to the same panel, as in **do**.

——————————————————————————————————

Plot similar lines in the same panel.

If there are no comparisons of particular interest, and no meaningful patterns in the configuration of content elements, then plot similar lines in the same panels, allowing them to be grouped perceptually into relatively few units.

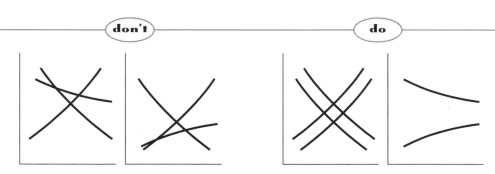

A judicious selection of data to be graphed in each panel may produce interactions that help tell your story.

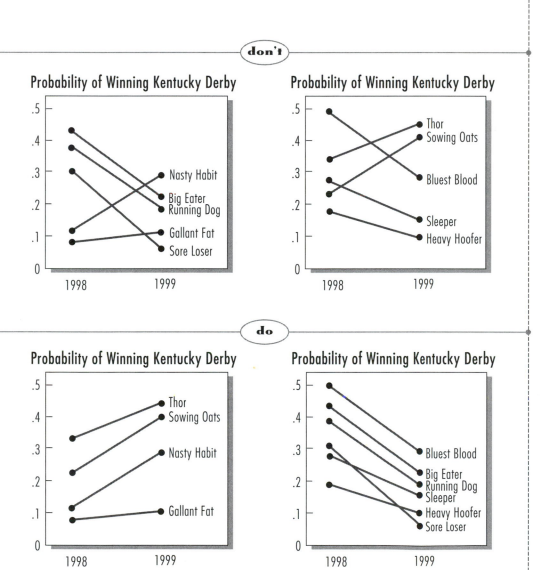

Picking a winner is difficult enough—the principles of perceptual organization can help.

Panels should be placed as close together as possible without causing improper grouping of components.

Multiple panels should cohere into a single display; if they are too far apart, they will appear as unrelated graphs. If they are too close, however, our visual systems may improperly pair content elements or labels of one display with those of the other (in accordance with the principle of proximity). The panels should be as close as possible without seeming to touch or otherwise running the risk of improper grouping of labels or segments. Compare the spacing of panels in **don't** and **do**.

Put the most important panel first.

If one of the panels presents information of primary importance, as in **do**, put it at the left or, if there is more than one row, at the upper left. Given our reading habits, this is the panel that will likely to be viewed first. If your audience does not read from left to right, modify this recommendation accordingly.

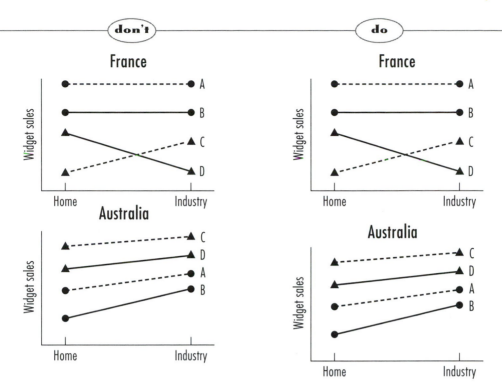

If panels are too close, labels and titles can easily be confused.

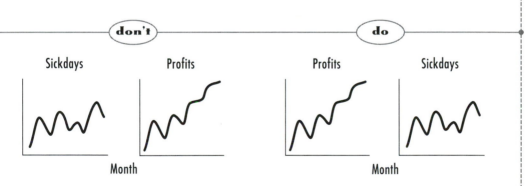

In most companies, profits are of more concern than sick days.

● **recommendation**

Overall heights in different panels should reflect overall amounts.

At first glance, from the **don't** version it seems as if all the countries spend roughly comparable amounts per pupil—but this is not true. The Y axes were excised so that significant differences could be seen between the pairs of bars in each panel. Now look at **do**, in which the data were plotted so that the overall differences are also apparent. When the Y axis of one or more panels is not continuous, the principle of compatibility implies that the scale on the Y axis should be constructed so that the relative ordering in total amounts is preserved by the relative ordering of the overall heights of the content elements, conveying the correct visual impression of ordinal amounts.

● **recommendation**

Corresponding content elements should have the same general appearance.

According to the principle of informative changes, all panels should be the same size and general appearance unless size or appearance is used to convey information: Two different-sized pies might illustrate the percentages of different types of weapons produced in each of two countries, the larger pie indicating that country produced more weapons overall. Or, a larger panel might be used to emphasize the main message. If such techniques are used, the title or caption should explain the point of the variation. Regions should be displayed in the same relative locations and in the same order in different panels, and regions that stand for the same thing should have the same appearance. If a dense cross-hatch is used for the "tanks" wedge of a pie graph illustrating the types of weapons sold by the United States, that pattern should be used for "tanks" in the pie graph illustrating types of weapons sold by France.

● **recommendation**

Use the same units along the Y axis in multiple panels.

When the same dependent variable is used in the different panels, use the same units along all Y axes, with the same number of ticks per interval. The principle of informative changes implies that using identical scales on the different frameworks generally is best, but the principle of relevance implies two caveats: If there are vast differences in the ranges, or only trend information is important, start the axes at different places. Be sure to indicate a break at the bottom of the X axis (by a zigzag or two slashed lines marking a gap) when the scale is excised.

don't

Government Spending per Pupil (1982-86 Average)

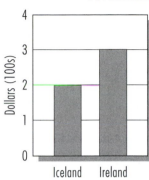
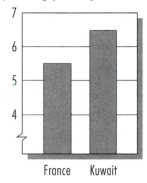
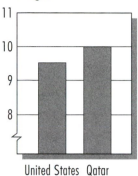

do

Government Spending per Pupil (1982-86 Average)

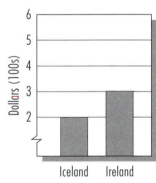
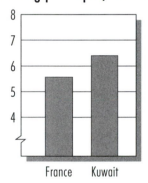
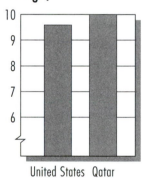

Arrange data to illustrate both pairwise comparisons and also relative overall levels.

Delete redundant labels on the axes.

Both versions of the example illustrate the percentage of people that in 1985 and 1986 believed that nuclear war and terrorism were the most important problems facing the United States. The **don't** display is the more cluttered version and contains redundant labels. If data are divided between two displays that are presented side by side, it often is possible to center a single label beneath both X axes. It may also be possible to eliminate the labels of the Y axis of the right panel, as is illustrated in **do**. However, it is important to keep in mind that eliminating the labels of the Y axis of the right panel makes it difficult to obtain precise values, even with an inner grid; this practice is appropriate only if the point is simply to illustrate trends and interactions under different conditions.

Specify part/whole relations explicitly.

The first panel of the example shows stock prices over most of 1990; the next panel shows the trend over the last week; and the last panel shows the trend over the last day of the last week. The arrows connecting the three panels are critical; they show how the panels interrelate. The titles of the individual panels are also essential. If a panel presents a second version of information in another panel, the relationship should be established by arrows or other visual means (for example, a drawing of a magnifying glass for a detailed subpart of a function or an exploded section). Without such cues, the proper grouping will not be established.

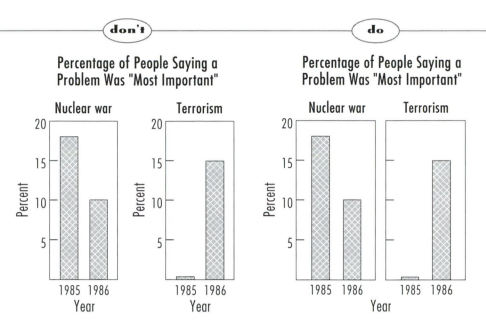

Every mark in a display competes for the reader's attention—eliminating unnecessary material is a good idea.

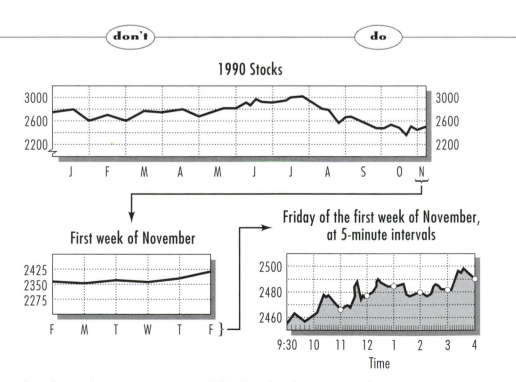

Complex graphs are sometimes unavoidable if much information must be presented. Indicating the relations among panels clearly helps the reader to comprehend the entire display.

In multiple side-by-side graphs, put labels of pairs at the far left and align the corresponding pairs in different panels.

When multiple side-by-side bar graphs are used, eliminate redundancy by using only a single set of labels, as in **do**. Because we read from left to right, it is best to put the labels at the far left so that they will be seen immediately. Ensure that the corresponding bars are aligned so that the good continuation serves to group each label with the corresponding pairs of bars in each panel.

**The
Next
Step**

If you have created a pie graph, divided-bar graph, or visual table, now go to Chapter 3 to label its elements appropriately and provide a title (pages 88–103). For all other types of graphs, go to Chapter 8. Many complex displays can only be created by an iterative process, by trial and adjustment. Sometimes the overall configuration of the display has unfortunate implications. An insignificant difference can appear visually dramatic or a real difference among measurements may not be obvious in the display. Chapter 8 provides recommendations that can be used to check how accurately the visual impression conveyed by a display in fact reflects the actual patterns in the data.

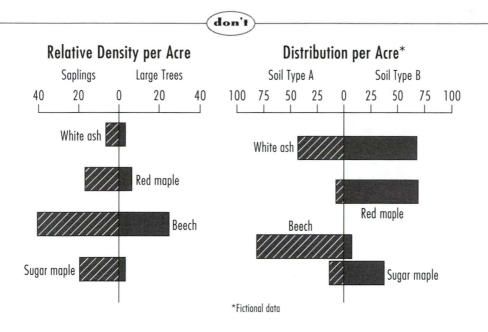

don't

Relative Density per Acre

Saplings   Large Trees

40   20   0   20   40

White ash

Red maple

Beech

Sugar maple

Distribution per Acre*

Soil Type A   Soil Type B

100 75 50 25 0 25 50 75 100

White ash

Red maple

Beech

Sugar maple

*Fictional data

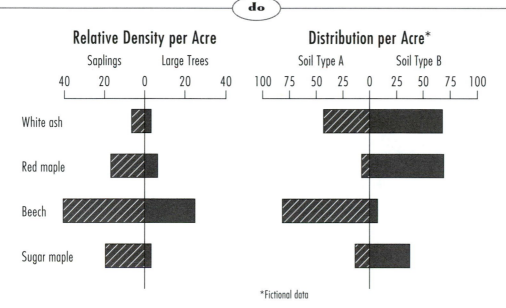

do

Relative Density per Acre

Saplings   Large Trees

40   20   0   20   40

White ash

Red maple

Beech

Sugar maple

Distribution per Acre*

Soil Type A   Soil Type B

100 75 50 25 0 25 50 75 100

White ash

Red maple

Beech

Sugar maple

*Fictional data

Take advantage of the principles of perceptual organization to reduce the number of labels.

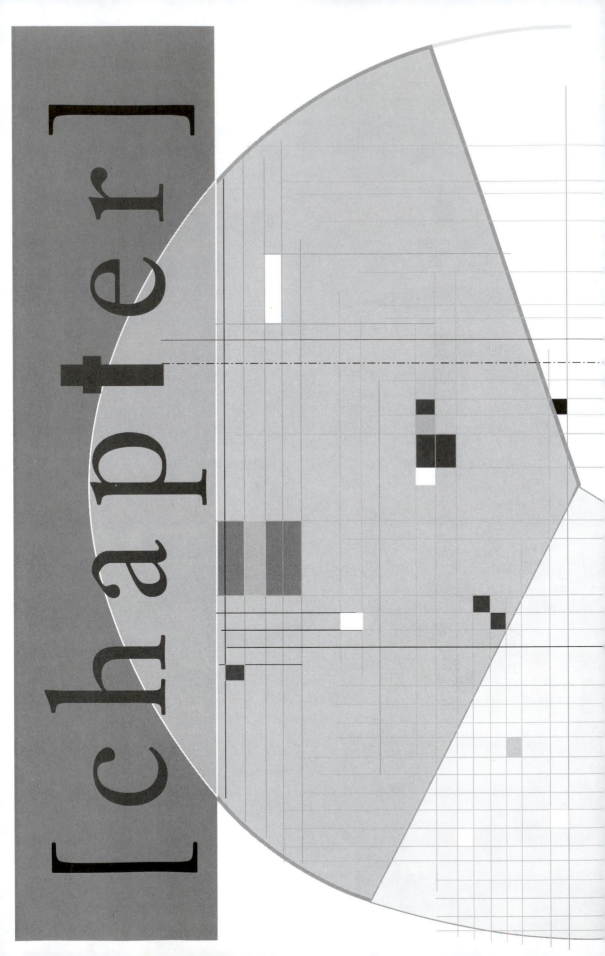

[chapter]

How People Lie
with Graphs

8

Visual displays allow us to gain a quick impression of relationships in a set of data. The principle of compatibility leads to a simple dictum for the display of quantitative information: More is (indicated by) more. Greater amounts of the measured substance should be depicted by greater visual amounts, either in extent (the height of a bar, point, or line) or in area (the size of a wedge in a pie graph or of a sector in a divided- or stacked-bar graph). In this way we can compare relative amounts at a glance; bigger visual differences are interpreted as reflecting bigger differences in the data.

This general idea is central to virtually all good graphical representations of quantitative data. As a designer of graphs you use it to invent new formats. As a reader of graphs, you use it to interpret a format that is new to you; simply look for the visual dimension or dimensions that are being varied and discover how those dimensions reflect variations in the content.

But our immediate apprehension of more-is-more is a two-edged sword. It makes it very easy for us to interpret accurately most types of graphical displays; it also enables a designer to deceive us easily, as was pointed out by Darrell Huff and Irving Geis in the classic book *How to Lie with Statistics*. To fool a reader, the designer merely needs to make the graph look as if there is more extent or area of one visual element than of another, and the reader will automatically assume that the actual entities represented vary in the same way. Many of the same techniques used to highlight important information or make actual differences visually obvious in a graph can also be used dishonestly to make spurious effects look significant.

This chapter is not intended to be a tutorial in how to produce dishonest graphs. To the contrary, my intention is to show the principles underlying such misrepresentation so that the reader will recognize a deceitful display for what it is. If enough people become aware of these principles, misguided designers may decide to stop using them to distort information.

The examples provided here do not exhaust the ways visual displays can be used to lie, but they convey underlying techniques that can be used: alterations of the framework, distortions of the content, and variations in the design of multiple panels. The distortions illustrated here range from the subtle to the blatant. Let us see how people can lie, and be lied to, with various types of displays.

Variations in the Framework

Visual differences in bar and line graphs can be exaggerated or minimized in various ways. Some of these methods require altering the axes. Consider, for example, this series of four graphs illustrating the numbers of strategic warheads for the United States and the U.S.S.R. in 1991. The first graph has not been distorted, and two things are apparent: First, both sides have a lot of warheads; second, there is only a small difference in the size of the two arsenals. Now, compare this graph with the next three, which might be used by (surely hypothetical) opposing political candidates.

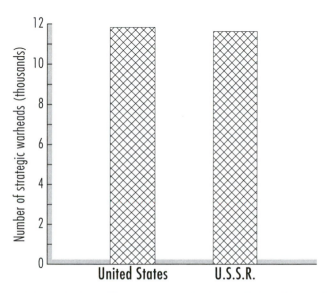

In 1991 the United States had 11,877 strategic warheads and the U.S.S.R. had 11,602. These data are presented in distorted form in the three following graphs. In order to focus attention on the competition between the United States and the U.S.S.R., in this series the labels for the two countries have been emphasized.

Do not excise the Y axis in order to exaggerate a difference.

Excising part of the Y axis is a useful way of making important variations in lines or bars noticeable. However, this technique, used improperly, can mislead the reader. The candidate arguing against a military buildup replotted the data on a graph whose vertical axis starts not at zero, but at 11,500. The difference in the number of warheads *looks* bigger on the page, and the reader is left with the impression that there in fact is a bigger difference in the stockpiles. Notice also that the excision is signaled by very small slash lines, which may not be detected by the casual reader.

Do not provide a spurious range of values to minimize a difference.

Another candidate, arguing for a military buildup, might want to minimize the difference between stocks of warheads and draw the graph differently. By extending the scale on the vertical axis from 0 to 1,000,000, the designer has made the *visual* difference in the number of warheads very small indeed. To make the added range seem necessary, a spurious reference line may be included, such as the number of warheads that would assure a total nuclear winter if all were detonated; this added information serves to distract the reader from recognizing the psychological effect of the range used.

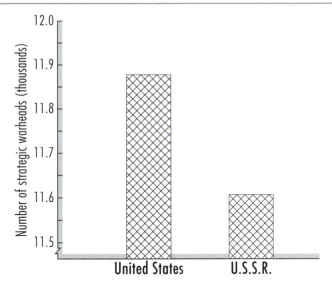

The data have been replotted using a Y axis with a restricted range of values (most of the scale has been excised), which greatly exaggerates the visual impression of the difference.

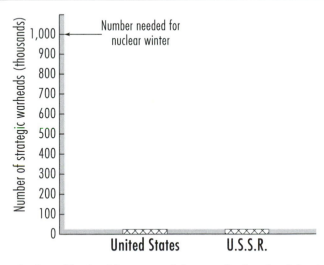

The data replotted using a Y axis with an expanded range of values (needed to include the number of warheads that would, by the most conservative estimates, ensure a total nuclear winter), which diminishes the visual impression both of the absolute numbers of warheads and the difference in numbers in the two countries.

Do not use a scale that creates a misleading impression.

Another way to minimize the visual impact of a difference is to use a logarithmic scale on the Y axis. In this sort of scale the distance between numbers becomes increasingly compressed as the numbers increase (see pages 80–81 and Appendix 1)—as the total amount increases, it takes larger differences in values to produce the same visual effect. This technique, used in the warheads example, erases any distinction between the stockpiles of the two nations.

Logarithmic scales are useful if the measurements have such a large range that critical differences among relatively small numbers are not visually evident—but the technique can be especially misleading if the content has two or more levels on a parameter (two or more lines, or two or more bars, per location on the X axis), as in the graph of stock-market growth. In this example, the top line appears to show a smaller increase in stock market growth in the United States than in the United Kingdom, illustrated by the lower line; in fact, the increase in the U.S. stock market was well over four times larger than that in the U.K. market.

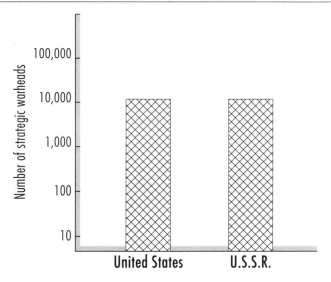

A logarithmic scale on the Y axis minimizes a visible difference if, as here, the overall amounts are relatively large.

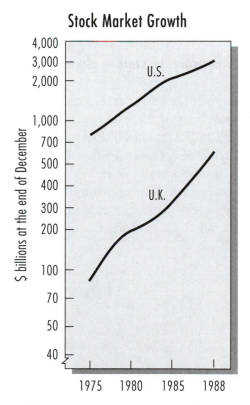

A logarithmic scale on the Y axis minimizes the apparent difference in trends because it reduces the visual impression of a trend when there are relatively large overall amounts.

Do not vary the aspect ratio to mislead.

We interpret a steeper line, or a sharper increase in the height of bars, as reflecting a larger increase in the measured quantity. A dealer trying to sell Asian elephants to a zoo might use **don't** to imply that Asian elephants are *considerably* smaller than African elephants and thus have *appreciably* lower upkeep costs. The **do** version presents the same data, but the difference by species is not nearly so dramatic because the slope of the line more accurately reflects the actual ratio between the average weights of the two species. These graphs show how variations in the aspect ratio—the ratio of height to width—can alter the visual impression of a difference. The effect of increasing or decreasing this ratio is to make the rise seem steeper or less precipitate. These visual impressions can be mistakenly translated by the reader as meaning a greater or lesser difference in the measured quantity.

Do not vary the starting places of bars.

The overall visual height of a bar or line conveys amount; the graph will be misleading if the baselines of these elements vary within a display. This effect is employed in "Saudi Arabian Oil Prices," where the graph is inventively drawn on the back of a camel. Prices did not in fact peak in 1986 (they were the same in 1986 and 1988), nor did they rise as sharply in earlier years as it appears here. The appropriateness of the camel illustration may distract the reader from noticing that the designer has added overall height to some bars by increasing the height of the base.

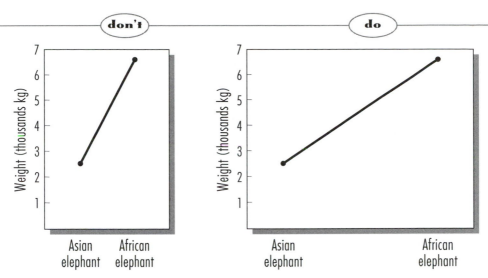

Increasing or decreasing the height of a display relative to the width can visually—and inaccurately—exaggerate a difference.

Saudi Arabian Oil Production

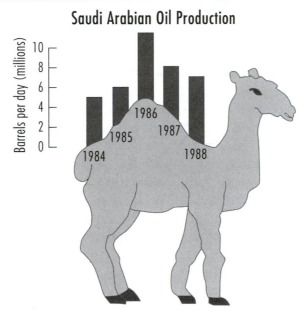

The lengths of the bars, not the heights of their tops, convey the information in this display (which is more like a visual table with a scale than a graph); because the baseline varies, the tops of the bars do not indicate the amounts.

215

Do not add a spurious reference line to alter perceived differences in height.

The designer can also include a reference line to alter the perceived origin. Now the reader sees the tops of the bars or lines relative to the heavy horizontal line. The example illustrates the 1985 energy consumption in tons of coal equivalent in the United Kingdom, West Germany, and Japan. The reference line is the energy consumption of Canada (which has nothing to do with the case); its inclusion visually amplifies the difference between the United Kingdom and Japan. The effect of a spurious reference line can be emphasized by shading or coloring the portions above the line, as was done here. Recall that we pay attention to differences in visual properties: We note proportions, not absolute amounts. Thus, by visually defining shorter bars (the portions above the reference line), the designer makes the differences in the heights of the tops more striking than if the entire lengths were compared.

Do not use a three-dimensional framework to exaggerate sizes.

Many displays are drawn in perspective, as if they were three-dimensional. This technique provides many opportunities, seized on by some designers, to fool the eye and the mind. The underlying phenomenon is known as size constancy, the tendency to perceive an object at its actual size at whatever distance it is seen, even though the object would occupy a smaller region in a photograph when it is farther away. A distant car does not look smaller than a near one, it just looks farther away. So, if two pictures of cars are drawn the same size on the page, but one is drawn so that it appears to be farther down the road, away from the viewer, that one will be seen as the larger of the two. By making the framework look three-dimensional, a designer can take advantage of this predilection to preserve the size of objects. Here this technique has been used to diminish the visual impression of the difference between Qatar and the United States in spending per pupil. The bars "farther away" look larger than they should because the visual system compensates for the apparent distance. Note that the baseline is different for each bar, which also contributes to the impression of increased height.

Energy Consumption, 1985

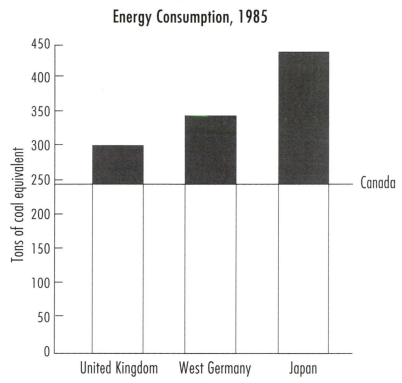

The eye notices the dark portions of the bars, which have greater proportional differences than do the bars overall—and thereby the reader is misled into seeing larger differences than are actually present.

Government Spending on Education per Pupil

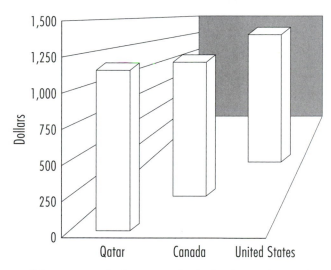

The phenomenon of size constancy leads us to see the farther bars as larger than they are, an impression that is reinforced by the added height on the page of the more distant bars.

Do not use perspective to change the scale at different locations.

A framework drawn in depth is sometimes used, perhaps inadvertently, to change the scale at different locations along the X axis. In "Oil Revenues" the graph is drawn on a drum, and the display would suggest that oil revenues peaked in 1985 and declined rapidly thereafter. Such a display might be included as part of a plea to raise prices. Because the central part has the most vertical range and both sides are compressed, this graph distorts the relative extent of the differences found in the middle years. The reverse effect will occur if the display is drawn on a surface that curves toward us (such as the inside of a coffee mug, seen from over one edge of the rim); when we look into a concave display, the middle portion appears the farthest from us. One sometimes sees graphs that are drawn on a wall that is receding into the distance; because the range of variation is expanded for the content on the "nearest" parts of the wall, these differences will be emphasized.

Variations in the Content

The data themselves can be transformed to produce an inappropriate visual impression. Special tricks can be used with specific sorts of content material, namely bars, lines, and regions.

Do not use percentages to exaggerate or minimize changes.

A particularly compelling example of the visual effects of transforming the data is illustrated in the two stock market graphs. "Size of Stock Market" tracks the *absolute* size of the stock market in three countries over three years. In the "Percent Change" graph the absolute amounts have been converted to percentage increases over the previous measurements. Note that the relative percentage of increase is larger for Japan than for other countries, a fact not immediately obvious in the first graph. If a designer wanted to emphasize Japan's economic strength, "Percent Change" could be used to conceal the fact that the United States actually had a much bigger stock market during these periods. It is no mystery which version would be used to sell U.S. bonds.

One can choose what percentage to graph: percent over all countries, percent relative to some baseline (overall average, lowest profits, highest profits), percent relative to the same country at an earlier time. Different visual effects will arise depending on what numbers are plotted, and numbers can be chosen that convey almost any desired effect. Percentages should not be used to obscure information the reader needs to make specific decisions.

Oil Revenues in the United Kingdom

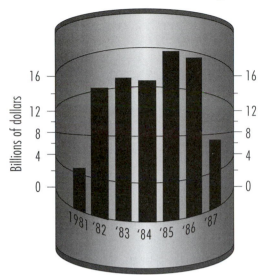

The scale at the middle is exaggerated compared to the scale at the sides, which exaggerates differences among bars near the center relative to those at the sides.

Size of Stock Market

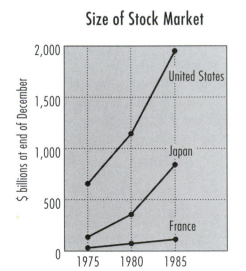

Percent Change in Size of Stock Market

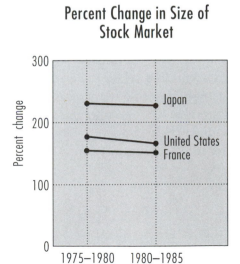

The United States stock market had larger increases per year and was much larger than Japan's (left); graphing percent change (right) obscures those facts.

• recommendation

Do not use derivatives to mislead.

Another transformation that can cause visual displays to tell one story when the data in fact tell another hinges on the use of the mathematical concepts of first and second derivatives. The first derivative is the rate of change, the second the rate of change of this rate: In the usual illustration, the first derivative of distance with respect to time is speed, the second is acceleration. One might graph the rate of population increase using a first derivative, which would indicate the increase from year to year in the number of people being born; the second derivative would indicate the *rate* at which the size of the increase changes. The second derivative can be very large even if there is only a modest increase in actual amount relative to a small base. Many readers, unfamiliar with derivatives, will interpret the bars or lines as displaying absolute amount. Even for mathematically aware readers, such a transformation can effectively obscure embarrassing differences (or lack of differences) in the absolute amounts.

• recommendation

Do not graph difference scores to mislead.

A closely related technique is to graph only differences between two or more measurements. In this case, the dishonesty arises when the designer chooses an inappropriate number to subtract from each of the amounts. The example shows the stock market data from page 219, plotted in a different form. Here the display presents the difference in the percent change from the two intervals (constructed by subtracting the left number from the right number plotted in the "Percent Change" graph for each country), and conveys a very different impression of the economic status of the three countries.

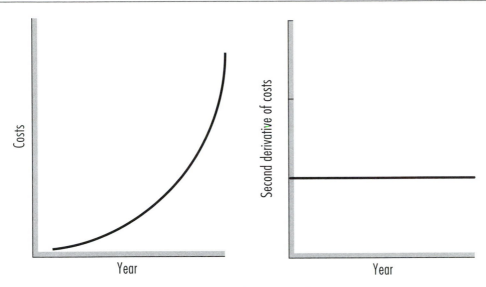

The graph at the right indicates that the rate of change in the rate of change was constant—but does not specify what the rate of change was. As is evident from the left-hand graph, very large rates of change can be constant.

Difference in Percent Change in Size of Stock Market, 1975–1980 vs 1980–1985

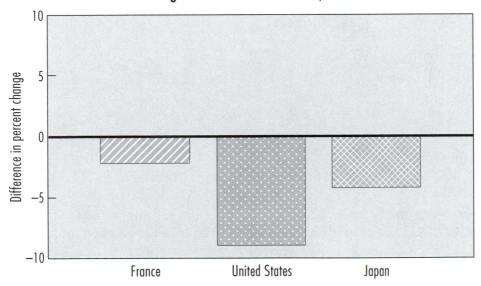

The reader has no way of knowing from this graph that the United States stock market in fact was the largest and fastest growing of the three.

Do not vary the thickness of wedges.

The "Maize Production" pie graph has been drawn so as to emphasize the growing maize production of China. The wedge for China has been exploded from the rest of the pie. It has also been drawn thicker, so its overall volume is larger—and it appears to represent more than it actually does. This effect depends on the principle of integral dimensions: You cannot pay attention to width without also paying attention to height.

Do not use three-dimensional edges to exaggerate the apparent size of the content.

Here the "Maize Production" pie graph has been drawn with a bit of a lip added in front to suggest a three-dimensional effect. This add-on carries the rim along quite nicely—increasing the visual impression of the United States' share of total maize production. The principle of good continuation groups the lip with the front wedges, making them seem larger than they are.

Do not use brightness to exaggerate area.

Compare the two versions of "Apple Production." Lighter areas are perceived as larger than darker ones (this effect, called irradiation, is a special case of the principle of perceptual distortion, and applies to brightnesses of color and, to a lesser extent, shades of gray). Using this variable to manipulate salience or to convey quantitative differences must be done with care to ensure that the display does not mislead.

Maize Production

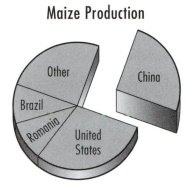

Making the exploded wedge thicker makes it seem to represent more.

Maize Production

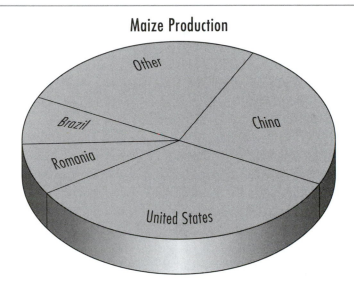

Adding the three-dimensional lip at the front exaggerates the size of the wedge for the United States.

don't   **do**

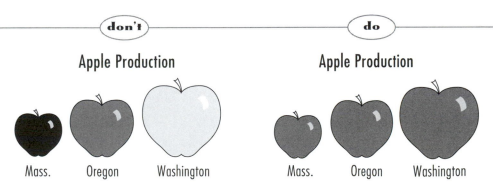

Adding intensity (left) exaggerates the visual impression of the magnitude of the differences in apple production.

Do not use areas to minimize differences.

In pie graphs and divided-bar graphs, the area of each sector conveys the relative amount. These displays can mislead if the visual impression of area is distorted. It is especially easy to be misled by such displays because our visual system is poor at estimating area; we systematically underestimate areas as they increase in size. So if one uses relative areas to convey amounts, one can de-emphasize an increase.

Do not covary height and width.

All three versions of the example purport to illustrate the brain weights of different animals. The top graph is not distorted visually, although for purposes of correlation of brain size with higher intelligence, the data themselves are misleading—the appropriate measure would be the ratio of brain weight to body weight, and cows would appear far less relatively sentient. The middle display might be used by someone who is opposed to the eating of red meat. This graph, by the selective use of a wider bar, seems to suggest that cows have larger brains than gorillas. The bottom display might be used by someone who favored meat eating. When subjects were asked to evaluate the rate of apparent increase from left to right, they rated the middle display as increasing more sharply than the top one, but the bottom display as increasing less sharply than the top one. Height and width are integral perceptual dimensions: Readers do not see the height or width per se, they see the size of the rectangle. This fact provides additional possibilities for lying with bar graphs and step graphs by varying the widths of the bars or steps as their height is varied.

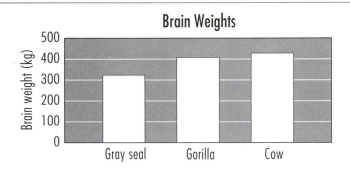

Brain Weights

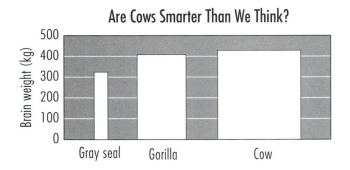

Are Cows Smarter Than We Think?

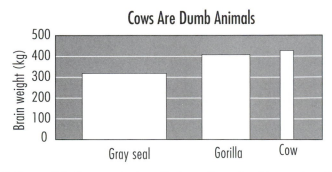

Cows Are Dumb Animals

Varying the thickness of the bar manipulates the immediate visual impression; the middle panel sends the message that cows are brainier than the other animals.

Do not vary the weight of shading, stripes, or cross-hatching to mislead.

The eye is drawn toward more visually salient regions. The person opposed to eating red meat might be tempted to create a graph like **don't** to draw the reader's eye toward the largest bar; making it more salient emphasizes its role in the display and leads the reader to overestimate the size of the increase.

Do not use occlusion to make bars look larger.

Tricks used to produce three-dimensional bars also can mislead the viewer. The example illustrates the percentage of people in each of five income groups. The rearmost bar represents 32.5% but looks much larger than that—a useful effect if the designer is trying to convince people to keep in office the incumbents who created so much prosperity. The eye sees partially covered objects as farther away than the objects that cover them. By having the bars partially overlap, the mechanisms of size constancy are brought into play, so the rearmost bars seem larger than they are.

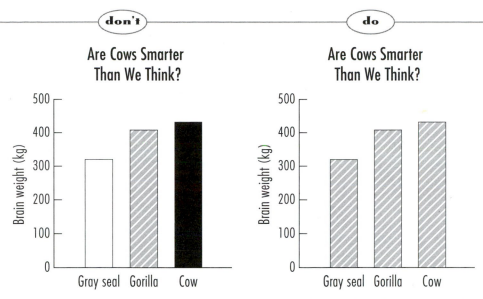

don't   do

The progression from gray seal to cow looks inappropriately sharper when additional ink makes each bar increasingly salient. If the background were dark, or if the display were on a computer screen, adding more lightness would have the same effect.

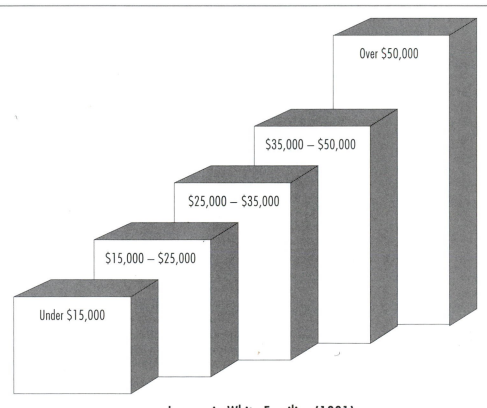

Income in White Families (1991)

We see occluded objects as behind the occluding ones, and size constancy operates to make the occluded objects appear larger than they otherwise would.

● **recommendation** ───

Do not vary width within a line.

The actual slope of "Average U.S. Airline Fares" is a line that bisects the
exhaust of the plane—but the *perceived* slope can be altered by shading the
top or bottom of the exhaust, as in **don't**. If the lines that form the content
of a line graph vary in width, the slope will be exaggerated if the top edge
is emphasized, or minimized if the bottom edge is visually stressed. This
technique looks silly unless the designer, as here, found a rationale for
decorating the content line.

● **recommendation** ───

Do not vary depth to distort line height.

If the content line in a line graph is drawn as a three-dimensional bar, one
end can appear closer than the other. The height above the Y axis (at the
"front" of the display) can be exaggerated if part of the line is drawn to
appear relatively far away. In the example, the increase in the German
share of world exports is exaggerated because the right portion of the bar
appears higher than it should. By having the bar snake back, the designer
makes the right part actually higher on the page—and the ambiguous
depth cues do not allow the reader to have a good sense of how high the
bar is above the "floor." This sort of exotic display is most convincing if
the line is drawn as an object (a bar of steel, in this case) that makes sense
in context.

Average U.S. Airline Fares

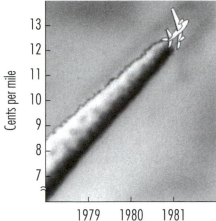

Average U.S. Airline Fares

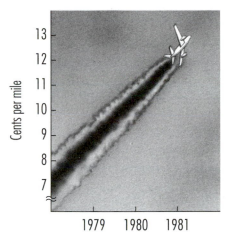

In the graph on the left shading has been used to emphasize the trend, making it seem steeper than it is. Shading the top would de-emphasize the trend.

German Share of World Exports

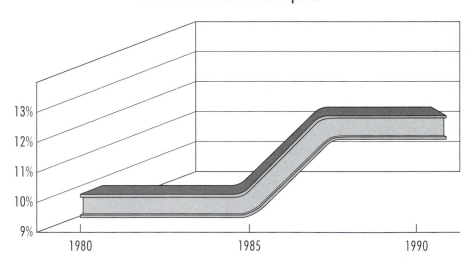

Varying the depth of the steel bar varied the height on the page, making it very difficult to read the actual values off the Y axes.

Do not insert a line in a scatter plot by eye.

When a cloud of points is presented in a scatter plot, fitting a line through them will enhance the visual relationship between values of the variables along the X and Y axes. In **do** there is in fact no systematic relationship between the X and Y variables, but by putting differently shaped lines through the cloud (flat, U, increasing), we get an impression of such a relationship. The effect occurs because of both the principle of proximity (points near the line are grouped with it) and the principle of common fate (points lining up with the line are grouped with it). This sort of manipulation only can be regarded as an intentional effort to mislead the viewer. The practice is particularly dishonest because there are well-established statistical techniques for fitting lines to data (see Appendix 1), and most readers will assume that such procedures were used to place the line.

Do not vary the aspect ratio to make dots in a scatter plot closer to the lines.

Varying the aspect ratio has the effect of making the points in a scatter plot seem nearer or farther from a line. When the width is relatively short, the points seem to fit the line better. The graph in the example shows the amount of time subjects in a study needed to scan across a visualized map with their eyes closed. Although it is clear in both versions that more time is required to scan a farther distance, the point is made more dramatically in the version on the right. This manipulation creates the impression of a stronger relation between the independent and dependent variables if the width is decreased, or a weaker relation if the width is increased.

don't   do

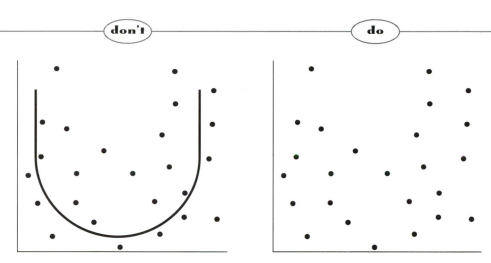

A line can impose a pattern on the points because they group with it (via proximity and common fate). As is evident in the graph on the right, there is no actual systematic pattern of variation in the plotted points.

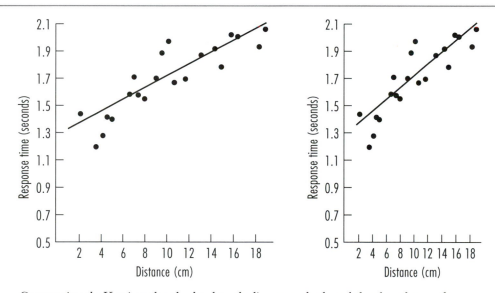

Compressing the X axis makes the dots hug the line more closely and thereby enhances the impression that they fit the line.

Variations in Multipanel Design

Another set of techniques that sometimes is used to mislead depends on breaking the display into two or more separate displays.

● **recommendation** ───

Do not vary the size of a panel to mislead.

The example shows the amount of time that first-graders, fourth-graders, and adults required to "see" large or small features of visualized animals (such as the back or the whiskers of a cat) versus the amount of time that they required to recall the same features without visualizing them.[9] The large panel shows that adults required different amounts of time to recall the information in the two ways. Note that one of the small panels shows a different trend: First-graders tend to use imagery even when they are not asked to do so. By drawing one panel of a multipanel display larger than the others, disparities can be swept under the rug.

● **recommendation** ───

Do not use different transformations or scales to distort the visual impression.

If there is an embarrassing difference between two sets of comparisons, the designer may try to conceal it by using different transformations (perhaps converting the data to logarithms in one panel and to difference scores in another). The reader will find it very difficult to make direct comparisons. Sales of the weakest division of a company might be shown as the percentage of increase, which is to be compared with actual sales for the other divisions. Of course, the designer would offer a rationale for this distinction, such as the relative youth of the weakest division and the claim that the amount of early growth is most important.

Time to Verify Properties of Objects

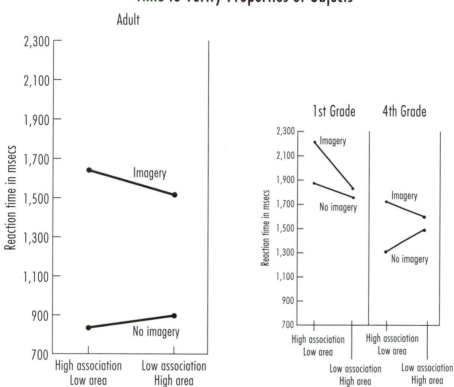

Adult

1st Grade   4th Grade

Making a panel larger and putting it on the left encourages the eye to linger on it—and thus possibly to neglect the inconsistencies to its right. This would be unfortunate here, because the difference for the first graders is what makes the graphs interesting.

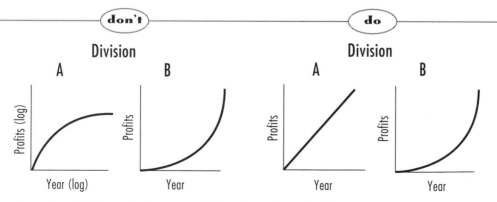

don't   **do**

Supporters of Division B might try to slip by the display on the left, arguing that Division A profits are about to peter out. The reality, shown on the right, is that both divisions are winners.

Do not combine distortions.

The three versions of "U.S. Exports to Mexico" illustrate a few of the ways in which techniques might be combined to produce even stronger effects. All visually exaggerate the amount of growth in the period 1989–91 and might be used to discourage Mexico from entering into a free-trade agreement with the United States. Research has shown that the techniques are not strictly additive: Combining two equally distorting techniques does not make the display twice as distorted psychologically. Nevertheless, such combinations do result in more striking—and thus misleading—displays.

The Next Step

We have seen how a simple psychological principle can be exploited in a variety of ways to distort the reader's impression of a graph's message. Readers should be alert for the use of such techniques (and the ones shown here are not the only ones); they are surprisingly pervasive in the media. You should take care not to use these techniques inadvertently to mislead the reader. If a pattern in the data is not significant, it should not be visually striking; if a difference or trend is obvious in a graph but is not actually significant in the data, the graph is misleading. If you find your graph is misleading, the next step is to correct the false impression by altering the framework or the content or both (without violating any of the relevant recommendations).

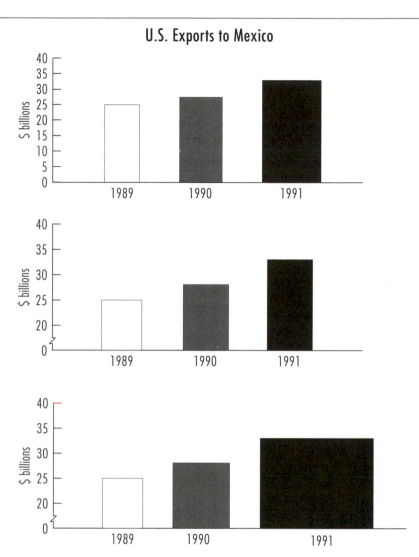

U.S. Exports to Mexico

Combining various techniques varies the degree to which a trend or difference is seen to increase or decrease.

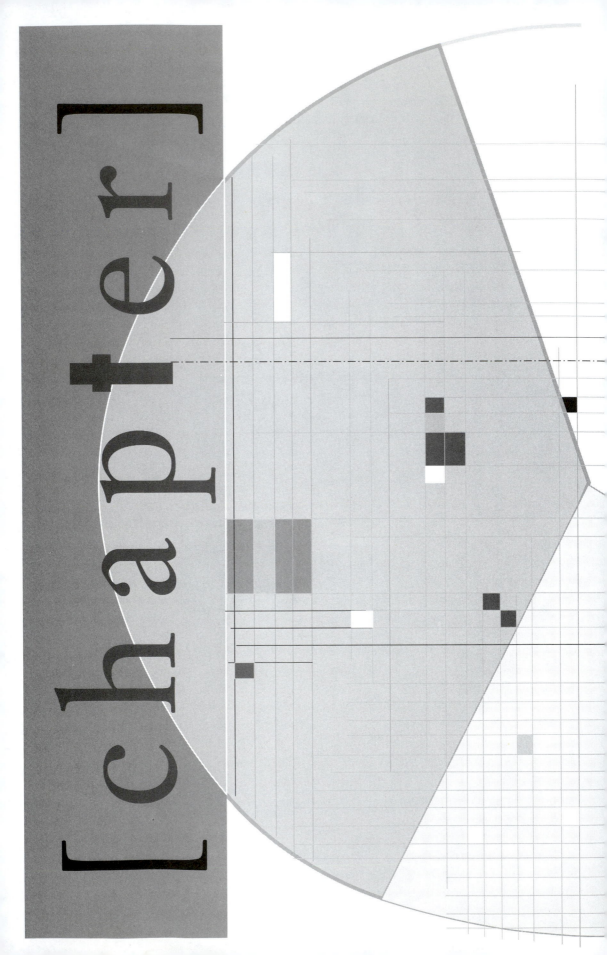

[chapter]

Beyond the Graph

9

The basis of the recommendations presented in this book is the recognition of a central fact about the reader, not the display: that the reader, as a human being, has a particular perceptual and cognitive system with identifiable strengths and weaknesses. That being so, the principles we have discussed will be in force whatever the specific nature of the display—the eye and mind, in their intimate partnership, will group elements, will ascribe meaning to visible differences, will in general behave as we have seen. The nature of the reader will not change; but the display the reader is looking at will not always be a graph. In this chapter I offer recommendations for using the principles of perception and cognition when creating other types of visual displays.

We can divide visual displays into two general classes, those that convey *quantitative* information and those that convey *qualitative* information. Graphs provide quantitative measures (say, the price of tea), even if the measurements apply to discrete entities (say, countries). Maps also fall into this category; as a consequence of portraying the layout of territory, they implicitly specify distances between locations, a quantitative variable. Often they also specify the amount of something at each location, perhaps height above sea level, population, or annual rainfall. However, maps cannot be used generally to convey quantitative information; they are tailor-made to convey information about spatial layout.

Charts and diagrams are qualitative. Charts specify qualitative relationships among entities; family trees and flow charts are kinds of charts. Diagrams are schematic pictures of objects or events that rely on conventionally defined symbols (such as arrows to indicate forces); "exploded" diagrams that show how parts of an object fit together and illustrations of football plays are diagrams.

In this chapter I offer some illustrative recommendations for charts, diagrams, and maps. These recommendations, by no means exhaustive, are meant to underline critical issues you should consider when designing these displays. They also show how the present approach can be generalized to a wide range of display types; it should be easy to see how the various principles (discriminability, salience, compatibility, etc.) can be applied to each type of display. Each of the following sections begins with a few comments about the circumstances in which you should use the format, and then turns to recommendations for designing that type of display.

Charts

By some definitions, a chart is considered to be any display that has a symbolic content—it stands for something other than what is directly pictured. This general characterization would include some maps (navigational charts) and graphs. I use the term here in a more restricted sense: Charts convey not amounts but relationships, not "How many work here?" but "Who works for whom?" Tables of organization, flow charts, and family trees are all charts, visual displays that arrange information into categories or structures. In graphs, the sizes of elements correspond to amounts of the entity being measured (which is why I refer to pie *graphs*). In charts, which do not measure, size may not correspond in any way to the entity depicted, and if it does, size will reflect not amount but a quality of the relationship such as subordination or inclusion. Usually, lines connect the elements of a chart to indicate structure; but the arrangement of the units (often boxes) on the page may be enough to show the relationship clearly.

● recommendation

Use a chart to convey overall organizational structure.

Using a chart to convey information about the organization of discrete entities takes advantage of the principle of compatibility. Relationships such as "is a member of," "follows," "works for," "is descended from" are illustrated clearly by the spatial relations of the entities in the display. A chart is a good idea if you want the reader to gain a sense of the overall organization of components, whether the management of a company, the degrees of kinship in a family, or the sequential steps in a process.

● recommendation

Use a description to convey few entities and relations, a chart to convey combinations of comparisons.

The alternative to a chart is a verbal description, which is preferable if only a few entities or relations need to be considered. A simple statement, "The company has two components, consulting and publishing," is better than a tree diagram. A short description probably can be read faster than a chart can be deciphered, and it will tax short-term memory less than would a multi-element display. On the other hand, if the reader needs to compare numerous different combinations of entities (such as the relations between various managers and their superiors in the company), a chart is better than a description.

● recommendation

More inclusive categories should be represented by symbols higher in the display.

The problem is to show how the Yamaguchita Corporation breaks down into separate divisions. The principle of compatibility leads us to place the larger entities higher in a chart with the entities that are contained within the larger entities depicted beneath them. Yamaguchita has automotive, consumer electronics, and ranching divisions. These would be symbolized in a chart by boxes (or pictures of representative objects) in a row directly beneath the box (or picture) representing the company. Beneath each of these three entities would be boxes or pictures for their components.

Use a layout compatible with the materials.

The principles of compatibility (pages 8 and 36–39) and cultural convention (page 38) apply to charts as well as to graphs. A chart detailing the
command structure of an organization should start at the top and work
down, as in **do**. By convention, a sequence over time progresses from left
to right, so a flow chart illustrating the steps of a manufacturing process
should start at the left and work to the right.

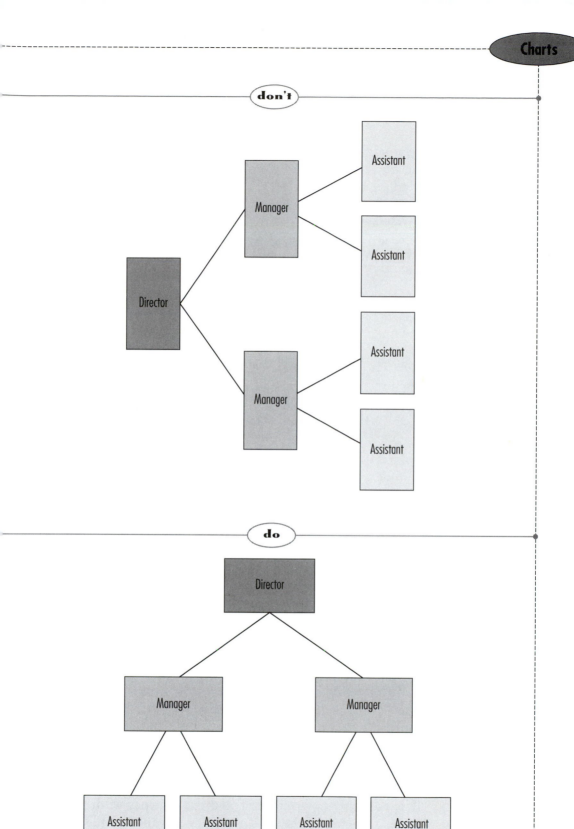

Category inclusion and command structure both correspond to vertical spatial relations; events over time are shown better in a horizontal layout.

Identify relationships

Remember the muddled display of nutritional information presented in Chapter 1 on page 12? Although the components of that display are two graphs and a table, the display as a whole functions as a chart. But, unhelpfully, the arrows leading from the center panel indicate different and unlabeled relationships. In keeping with the principle of relevance, identify every important piece of information in a display, as in the **do** version of this family tree.

Use different colors, shading, or line weights to organize the components of complex charts.

The **don't** version is difficult to take in; **do** is clearer. A chart is useful in conveying a complex set of qualitative relations, but the complexity must not visually overwhelm the relations depicted. This can happen if the number of entities or the number of relations among them is large. The spatial organization of the display, if properly produced, will allow the principles of perceptual organization to reduce the load on processing. To avoid producing a tangled web, divide the relations into two or more types and use different colors, shadings, or line weights for each: In **do**, the phases of the data collection process are distinguished from the evaluation and decision processes. If the content material does not lend itself to such a division, highlight what you consider to be the most important components. In some cases, however, even this will not help, and you should break the display into two or more individual displays.

Use familiar formats.

Charts can depict many types of qualitative relations and can do so in many ways. It is tempting to invent new and exotic types of display designs. This is well and good if the reader can quickly discern the entities and relations, but as any reader of national news magazines knows, this is not always the case. To be on the safe side, be sure that the reader is familiar with the conventions and types of display you are using.

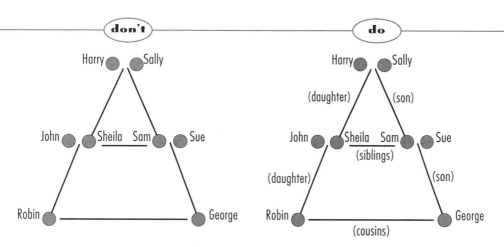

Every important piece of information, whether entity or relationship, should be identified in the display.

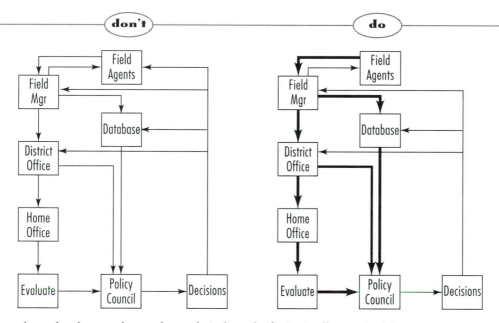

A complex chart can be sorted out relatively easily if it is visually organized into components.

Diagrams

Diagrams are pictures of objects or events that use conventionally defined symbols to convey information—to show the wiring of your kitchen, the nitrogen cycle, or the assembly of your model 1968 Mustang. Diagrams combine literal elements (pictures of parts) and symbolic ones (arrows to show movement, direction, or association; shading to show curvature).

● recommendation ————————————————————————————

Do not explode an object too widely for recognition.

The brain processes shape and spatial relations in separate systems and does not combine the two very accurately (recall the principle of imprecision). Do not expect the reader to extrapolate or remember precise spatial relations among parts of a display. In an exploded diagram, show parts near their actual locations on or in the object depicted, as in **do**.

● recommendation ————————————————————————————

Include only relevant material and ensure that it is visually salient.

Always respect the principles of relevance and salience: The most important material should be highlighted, and unimportant material de-emphasized or deleted. A diagram meant to show the driver's controls in an automobile may not be helpful to the reader if all automotive systems are drawn in, as in **don't**.

● recommendation ————————————————————————————

Make conventions clear.

Certain disciplines—electronics, genetics, and linguistics, to take a few examples—have their own vocabularies of symbols, which should be respected. Depending on context, however, the same symbol may have different meanings: An arrow may indicate parts to be matched, direction of force, or movement. If you think your audience may be in any doubt about the meanings of the symbols you are using, state those meanings in a caption.

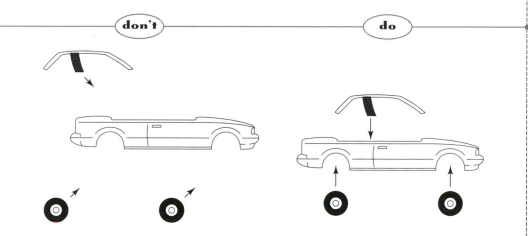

If spatial relations are distorted to show components, the distortion should be easy to reintegrate; otherwise, readers will have to expend effort to see the components in place.

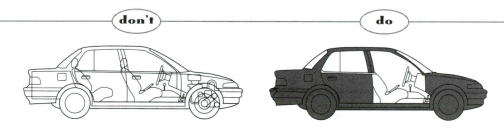

A diagram intended to show the driver's controls is easier to read if the irrelevant components are eliminated and the relevant ones highlighted.

Maps

Maps are drawings that function as pictures of a physical layout: The features of a room, your town, the earth, the sea, and the sky can all be mapped. As stylized pictures of a territory, they provide information about locations and relations among them (usually in terms of relative distances and routes between them); by using conventional markings, they can also provide quantitative information about various locations such as average temperature, voting patterns, population distribution, and so on.

● **recommendation** ──

Use a map if more than one route is possible.

One alternative to a map is a verbal description of a territory. If the reader can reach the destination from a number of different starting places or by more than one route between two locations, a map is distinctly more helpful (remember the last time you asked for directions). A map is also preferable if the reader is likely to want information about the distance or spatial relations among numerous locations. In these circumstances a map will reduce the load on short-term memory.

● **recommendation** ──

Avoid visual illusions that distort distance and direction.

Visual illusions affect our perception of distances. Vertical lines appear longer than horizontal ones (see pages 114–115), and straight lines that are interrupted, as in **don't**, appear to be displaced; the diagonal road passing under the highway is actually straight, but it appears to jog under the overpass. This is not a problem in **do**.

● **recommendation** ──

Use a map to label complex sets of information about a territory.

Maps are also a convenient way of labeling information about localities. A map allows the reader to see the relations among different values in many locations. For example, if population is presented (perhaps by bars standing on particular locations), the reader can see not only the relation of population to specific landmarks, such as rivers or the sea, but also gain an impression of the pattern of population variations as a whole. A map allows symbolic content to be grouped with locations via the principle of proximity.

don't   do

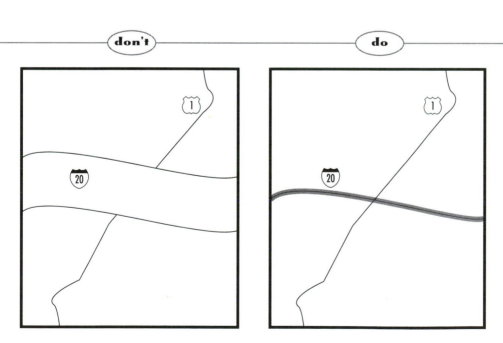

Beware of visual illusions (including exaggerated distances of vertical lines) when producing maps.
In the map on the left, Route 1 seems to jog as it passes under the Interstate, but it does not.

● **recommendation**

Do not vary height and width of location markers to specify different variables.

The map in **don't** uses the height of the markers to indicate the population and the width to indicate the mean temperature of each location. But the eye sees the area of the markers, not each dimension separately. Do not vary integral dimensions (such as height and width, or hue and saturation) to specify values of distinct variables (see pages 116–117).

● **recommendation**

Do not vary area of regions to convey precise quantitative information.

Some years ago it became fashionable to design maps of the United States in which the sizes of the states were varied, as in **don't**, to show relative rates or quantities—per capita beer consumption, number of murders, and so forth; the more per the state, the larger the area. Such maps can convey a rough impression of rank ordering, but fall short if actual amounts or even precise ordering is to be conveyed. We are simply not very good at estimating area and have trouble comparing relative areas of differently shaped regions. A better way to convey amount is to draw a bar or similar symbol at each location; our visual system can compare relative line lengths well.

● **recommendation**

Ensure that regions are identifiable.

Another problem with varying the size of a region to convey information is that the shape may become unrecognizable. If Maine is made huge, its shape may have to be modified to fit it against its neighbors, as happened in the **don't** example above.

don't   **do**

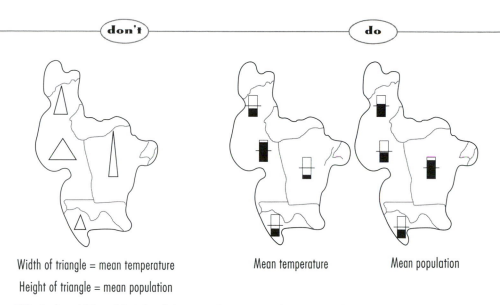

Width of triangle = mean temperature   Mean temperature   Mean population

Height of triangle = mean population

If both the width and height of the triangles are varied, we see neither variable very well; it is far better to use two different displays, one for each variable.

don't   **do**

Projected Snowfall, 1995   ## Projected Snowfall, 1995

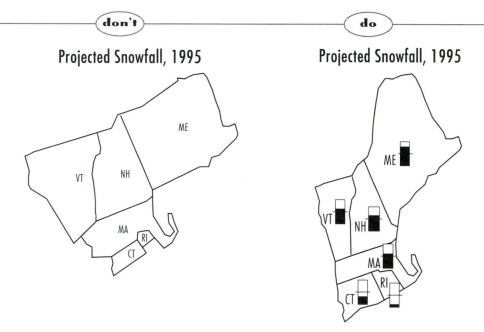

It is almost impossible to extract relative amounts from differently shaped areas, even if they are familiar and not distorted. Vary extent along a line or bar instead.

• recommendation ───

Provide neither more nor less detail than required for the purpose.

The principle of relevance is often violated in maps. If you want to illustrate how to get to the fire station from Harvard Square, it would not be helpful to include every road in the region, as in **don't**. Instead, illustrate the roads that are possible routes between the two locations, providing enough landmarks that the traveler won't get lost.

• recommendation ───

If distance is important, use grid markings.

Some maps are intended to indicate the relative locations of regions, the locations of specific routes, or to provide other qualitative kinds of information. Other maps, however, specify the actual distances between locations. In these cases, inner grids are useful and should be drawn following the recommendations presented on pages 182–185: Inner grid lines should be relatively thin and light (this recommendation is violated in **don't**); use more tightly spaced grid lines when greater precision is required; insert heavier grid lines at equal intervals; inner grid lines should pass behind the lines, bars, or other symbols; the scale should correspond to increments of the grid.

don't   **do**

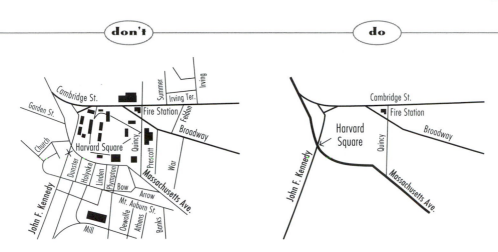

A map that includes too much extraneous detail for its purpose is not helpful; to specify the location of the firehouse, only the main routes are necessary.

don't   **do**

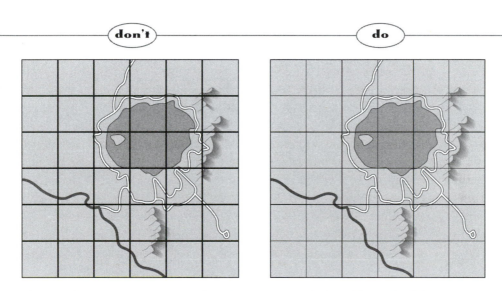

If precise distance is not important, there is no reason to include grid lines; if grid lines are included, they should not obscure the features of the territory.

The scale should equal one or more grid units.

The scale of a map is a kind of key, pairing a line length with a unit of distance, which the reader uses to estimate the distances between the locations represented on the map. To be effective, the scale should correspond to one or more of the units used to produce the grid lines. If it does not, the reader will have to reorganize the unit line length, mentally dividing it into the corresponding lengths on the grid; this operation is neither efficient nor accurate. If the grid lines are spaced every half inch on the map, the scale should indicate the distance in terms of a familiar multiple of this increment, say, 1 inch. Ideally the scale should indicate a unit that corresponds to the distance between heavy grid lines, as in **do**.

Make more important routes more salient.

Use the most important routes as the backbone of the map, as in **do**, allowing other routes to be organized by them. This can be accomplished simply by making the more important routes more visually distinctive.

Label important distances directly.

If you want the reader to know the distances between specific pairs of locations, draw lines between them and label them directly (the lines should indicate routes); this recommendation follows from the principles of relevance and informative changes.

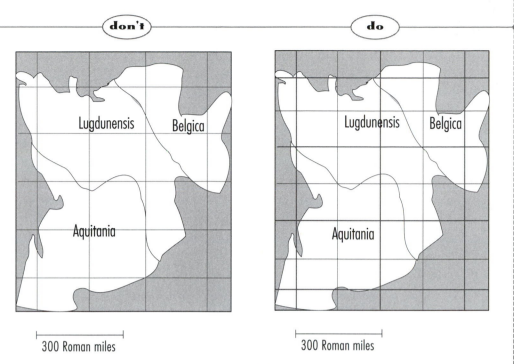

If the scale does not line up with the increments of the grids, mental division is necessary to use it; spare the reader this effort by aligning the scale and grid increments.

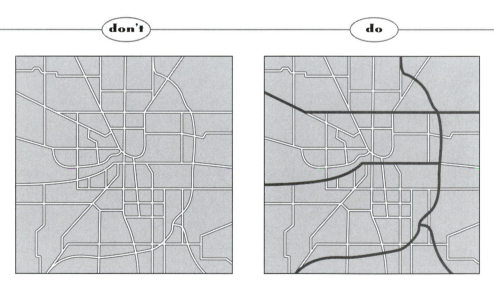

If the more salient lines correspond to major roads, the reader is helped to organize the map as well as to find the most efficient routes.

253

Into the Future

Perhaps the most exciting extensions of the psychological approach to graph design take advantage of new technologies. Computer-aided design (CAD) systems allow displays to move in various ways, and motion is another cue human beings use to organize and discriminate among objects. The design of CAD systems can be made "psychologically optimal" more easily than can charts and graphs because the designer can take full advantage of the dynamic aspects of the computer. Parts of the display can be selected for viewing as needed, with only the relevant information screened at any one time. If you are designing the layout of the electrical system for a new plant, it may be useful from time to time to compare this design with the layout of the plumbing system; but you would not want the two systems always displayed together as you work. Indeed, at one time you may want only to compare one aspect of the plumbing, say, the main sewage trunks, with the layout of the electrical system; at another time, you may want to compare some other aspect of the two systems. In CAD systems the amount of available information can be vast, without overloading our limited-capacity memories.

The design of such dynamic interactive systems would also benefit from the sort of cognitive engineering employed in formulating the recommendations in this book. If we can work out the principles that underlie the "psychological organization" of the object to be displayed—building, car, whatever—we can design the system to take advantage of the way people would break it down into regions and subsystems. These principles can be used to determine the ways in which specific information is added and deleted from the display.

By understanding the psychology of human perception and cognition, a designer is well placed to take advantage of readers' strengths and avoid falling prey to their weaknesses. The process of continuing to work to acquire this understanding is of great importance, even if no immediate application is obvious. Much pure research produced results that helped us solve the very practical problems addressed in this book. As illustrated here, such research often turns out to be of more than "academic interest."

appendix 1
Elementary Statistics for Graphs

R. S. Rosenberg and S. M. Kosslyn

This appendix provides a brief review of statistical concepts that are used in this book.

Variables

A *variable* is a measurable characteristic of a substance, quantity, or entity. The simplest form of a quantitative relationship measures one variable against another: heat in degrees against time, sales in dollars against season, tax revenue against each of several cities. Looking at these pairings, we can see that the values of one variable will change in relation to the values of the other: How hot the soup is depends (up to a point) on how long it's been on the burner; the sales depend on the time of year; the tax revenue depends on whether the city is Los Angeles or Toledo. Therefore one of the variables—the quantity whose changes we want to observe—is called the *dependent variable*; the other—as it were, the measuring stick—is the *independent variable*. In a line graph, for example, the dependent variable is graphed on the Y axis and the independent variable is graphed on the X axis.

But look again at the examples. "Time," "temperature," and "revenue" are not quite like "city." Time, temperature, and revenue are *continuous* variables—the values are numbers that change from one into the other. You can perform mathematical operations on these numbers, adding or subtracting two values, multiplying them, and so forth. In contrast, "city" is a *categorical* variable. It does vary—it can be either Los Angeles or Toledo; but it does not vary continuously, that is, it does not change from one into the other, and you can't perform mathematical operations on these values. Commonly used categorical variables are gender, location, team, and political party.

Scales

Variables are measured along several different types of scales; traditionally, there are four types, each with its own properties and arithmetic restrictions. A *nominal scale* (also called a categorical scale) presents values of categorical variables, names of the individuals, groups, or categories for which data will be shown. There may be numbers assigned to such a scale—the uniform numbers of the players on a team, or the numbers assigned to television channels—but these numbers function as names and cannot be manipulated arithmetically. This kind of scale may be arranged with a sense of number (for example, you may choose to arrange the cities for which you are plotting data from smallest to largest), but the members of the scale are discrete and the scale is in no sense continuous.

An *ordinal scale* arranges data by rank, ordering it into first, second, third, and so forth. Candidates for political office could be ranked by their standings in opinion polls, runners by their finish places. Rank is all that an ordinal scale shows: You cannot assume that the first-place candidate or runner is as far ahead of the second as the second is of the third—that is, you cannot assume equal intervals between the ranks. Only the ordering of the entities is important.

To know and compare intervals, you need an *interval* scale or its close cousin, a *ratio* scale. Both measure equal distances along a continuous scale, and the distinction between them is subtle. On a ratio scale, there is a set starting point from which the quantities are measured, an absolute zero, and so meaningful ratios can be computed from the intervals: Two weeks is twice as long as one week, ten pesos is twice as much as five pesos, and—on the Kelvin scale, which begins at absolute zero, the "starting point" of heat—40 degrees is twice as hot as 20 degrees. The last example pinpoints the difference between ratio and interval scales: When temperature is measured on the centigrade or Fahrenheit scale, neither of which is calibrated from absolute zero, 40 degrees is *not* twice as hot as 20 degrees, merely 20 degrees hotter. Data on a ratio scale permit more arithmetical and statistical operations than do data on the other three scales. For purposes of graphing data, however, the distinction between ratio and interval scales is usually not important, and indeed it has not been made in the discussions in this book.

Expressing Values

Depending on how data are construed, different information will be revealed. It is often useful to know the *central tendency* of a set of numbers, the locus of the most typical values, or *scores*, for a given sample or group.

There are three common ways to express central tendency. The most frequently used is the *mean*, or average, which is obtained by adding all the scores in a group and dividing the sum by the total number of scores: using the values in the table, $(7 + 1 + 2 + 8 + 1 + 4 + 5)/7 = 4.0$. (The 0 in 4.0 specifies that the result of the division is accurate to the nearest tenth.) The *median* is the value at the midpoint of a series of scores—half the total

Daily Production of Widgets at Seven Factories

Factory	Number of Widgets	Squared Deviations
A	7	$(7 - 4)^2 = 3.0^2 = 9.0$
B	1	$(1 - 4)^2 = -3.0^2 = 9.0$
C	2	$(2 - 4)^2 = -2.0^2 = 4.0$
D	8	$(8 - 4)^2 = 4.0^2 = 16.0$
E	1	$(1 - 4)^2 = -3.0^2 = 9.0$
F	4	$(4 - 4)^2 = 0.0^2 = 0$
G	5	$(5 - 4)^2 = 1.0^2 = 1.0$

Cases = 7   Total = 28   48.0 = Sum of squares (SS)

Mean: Total number/number of cases = 28/7 = 4.0
Median: Value representing 50th percentile = 4
Mode: Most frequently occurring value = 1
Range: Largest value – smallest value = 8 – 1 = 7
Variance (corrected): Sum of squares/(number of cases – 1) = 48.0/6 = 8.00
Standard deviation (corrected): Square root of variance $= \sqrt{8} = 2.83$
Standard error of the mean (corrected): Standard deviation/square root of
          (number of cases – 1) $= 2.83/\sqrt{6} = 2.83/2.45 = 1.16$

number of scores fall above that value, half fall below. Arranging the same numbers as 1 1 2 4 5 7 8, we see that the median is 4. The mode, the value that occurs most frequently in the sample, in our example is 1. The mode may fall at the high end or the low end of the series, or anywhere along it. In most cases, however, the mode usually falls on or near the other two measures. The mean, median, and mode can be computed on ordinal, interval, and ratio scales. The mode can also be computed for nominal scales, but it must be stated slightly differently—the "most frequent value" on a nominal scale might be read as "most [of the group] are women" or "most [of the group] are third-graders." The mean is the most sensitive to extreme values, the mode the least sensitive. In general, a median is more appropriate than a mean if the data are not distributed along the familiar bell-shaped curve (called a *normal* distribution), on which most values fall in the middle range and the extremes taper off in symmetrical tails.

If scores are presented directly, not expressed as measures of central tendency, they are *raw data*. They may be transformed in various ways to provide more information, indicating where they fall relative to one another. A *percentile* rank of a given score specifies the percentage of cases scoring at or below that score. Converting or transforming a score to a percentile rank immediately conveys where that score falls compared to the rest of the group or data set. Quartiles divide the group into fourths—the 25th, 50th, 75th and 100th percentiles—so a score at the first quartile (or the 25th percentile) signifies that 25% of the sample fall at or below that score. Similarly, deciles divide the group into tenths: A score at the third decile (30th percentile) signifies that 30% of the sample fall at or below that score.

Another method of transforming scales is by using logarithms. This is a useful technique if the scores have extreme values and important variations occur among small values. The logarithm, usually based on 10 or 2, indicates the amount that the base value would have to be raised to produce a given number. If base 10 is used, the log of 10 is 1, the log of 100 is 2, and so forth. The amount that 10 would be raised is specified by the exponent. If 10 is raised by a factor of 2, the exponent is 2: $10^2 = 100$. Ten is a convenient base because the exponent indicates how many zeros follow the digit 1. Fractions of an exponent can be used, so all numerical values can be expressed. If logarithms are plotted, the equivalent distance on a scale represents increasing amounts as you ascend the axis. This is of interest in creating graphs because by using a log scale equivalent distances on the axis represent progressively increasing amounts: The first interval represents 10 units, the next 10^2 or 100 units, the next 10^3 or 1,000 units. Differences among large numbers are compressed, because the same distance on the scale stands for increasingly large increments as numbers get larger. As a consequence, more of the Y axis is available to display variations among small values than would be available if the untransformed values were plotted.

Measures of Variability

Variability conveys information about the spread or scatter of the scores. The simplest measure of variability is the *range*, which is obtained by subtracting the smallest score or value in the set from the largest score or value. Like the mean, the range is sensitive to extreme scores and does not give information about how the variability is distributed over the scores. Two factories may have the same range in annual income of managers, but most of the managers at factory A are at the lower end of the pay scale whereas the salaries of managers at factory B are scattered equally throughout the pay scale. Expressing the range would not capture this difference.

You can also express variability by presenting information about the range of some proportion of the data. For example, if medians are graphed, you could use I-shaped brackets (as illustrated on page 59) to plot the values of scores that are 20% of the way toward the extreme values. This practice will be most effective, however, if readers have some idea of the stability of such measures.

Perhaps the most common method of expressing variability is the *standard deviation*, which describes the average variability for a group or set. It is based on the average of the deviations of each score around the mean, which are called the deviation scores. More specifically, because the sum of the difference of each score from the mean will always equal zero, the standard deviation is computed by squaring all the deviation scores and then summing them; this is called the sum of squares (or, in most statistics textbooks, SS). That sum divided by the number of deviation scores used to obtain it is the *variance*. However, to be a useful measure of variability that can be associated with a measure of central tendency, we must take the square root of the variance (remember, we squared the deviation scores at the beginning so we must "undo" the square); this square root is the standard deviation.

There are two standard ways to estimate the variance. The one just described is sometimes used in summaries of central tendency and variability. But if you want the reader to be able to decide whether the difference between two means is statistically significant (reflects an actual difference between the measured entities, rather than an accident of how the measurements were obtained), the variance must be computed by dividing not the total number of cases that went into it but that number minus 1. (This is a truer measure because the variance from a sample is an underestimate of that from the entire population; subtracting 1 from the total number of cases is a correction factor, producing an unbiased estimate of the variance.) The square root of this variance can be taken in turn to produce a standard deviation. This measure is appropriate to plot in a graph.

Another measure of variability is the *standard error of the mean*, which reflects the variability of means of your sample. Imagine taking a set of 10 apples at random from a supermarket bin, weighing them, and then taking the mean. Then take another set of 10 apples, weigh them, and take their mean. The two means would not be exactly the same. If you took a third sample, its mean would probably be a little different from the other two, and so on. You can estimate the variability of these means simply by taking the standard deviation of one sample (one set of 10 apples) and dividing it by the square root of the number of cases (10, in the example). You can then plot one standard error of the mean above and below a point on a graph.

It is a good idea to plot the standard error of the mean on graphs because it is a comment on the data you are presenting. The reader sees means, indicated by points or dots, and wants to know how stable they are—that is, how seriously to take them. The standard error of the mean indicates the range in which one could expect to find the mean if repeated samples were taken.

259

Best-Fitting Lines

A scatter plot is often most useful if the trend in the data is highlighted by a line fitting through the cloud of points. Depending on the relationship between the variables being plotted, this may be a straight line or one of several curves. The line is, in a sense, the ideal representation of the trend: If no data point deviated from the trend, all the points would lie on the line.

But they don't. The problem is to lay the line along a path from which the data points deviate the least. The most common way of doing this is by the *method of least squares*. The vertical deviation of each point, whether above or below the line, is squared (to eliminate positive and negative values that would cancel each other out) and the squares are summed. The line is in fact an ideal representation of the trend—that it, it is best fitted—when the sum of the squares is the smallest possible. Needless to say, a trial-and-error approach to the method of least squares is a formidable prospect, and in practice the line is placed by algebraic formula, and a calculator with statistical functions (or a computer with a suitable graphs program) is of inestimable value.

A related concept is the *correlation*, usually expressed by r, which provides a measure by which to judge how well measurements on one scale are predicted by measurements on the other. If $r = 1.0$, the points would fall perfectly on a best-fitting line; if $r = 0$, the relationship would be random. Again, this value is found by using algebra. The correlation value squared tells you what percentage of the variability in the values on one scale are accounted for by the variability in the values on the other scale. For example, if $r = .80$, then $r^2 = .64$, and 64% of the variation in the values of the data points can be explained by the relationship between the two variables. Correlations are symmetrical; values of either variable can be used to predict values on the other.

Lying with Statistics

Graphs can be misleading in part because misleading numbers are presented (as was shown brilliantly by Darrell Huff in *How to Lie with Statistics*). Different statistics or comparisons represent different facets of the data. Suppose you are the public relations expert for a political candidate; part of your job is to issue press releases announcing the latest standing of your candidate in the polls. There are many ways you could do this. You could report the actual number (absolute value), or the percentage of people who favored your candidate. You could compare this number (or percentage) with many other numbers, depending upon the point you wanted to illustrate. You could compare it to your opponent's standing, your candidate's standing last month, to the projections from last month, to the projection of how many votes (or what percentage) are needed to win, to your candidate's all-time low or all-time high standing. You could report not the actual number or percentage of those people favoring your candidate, but the percentage increase. This would be especially useful if your candidate, who has been trailing in the polls, increased from 5% to 10% in share of the projected vote: You can state that your candidate's popularity has doubled. You could report your candidate's mean standing in polls, the median, or the mode. Each choice illustrates a different point.

As you can see, even before you start graphing the data, the statistics that you choose to represent the data create a bias in how they will look and the message the reader will receive. Be a skeptical reader and ask yourself this question: If the data were presented using different statistics, or compared to a different referent, would the message be different?

appendix 2
Analyzing Graphics Programs

New or revised graphics programs for personal computers are released practically every other month, and it would be silly to try to provide a comprehensive review of all graphics programs available. It would be equally silly to ignore them and not formulate a means of review. The checklist below is intended to provide a structured way to evaluate graphics programs and choose a package suitable for your particular needs. It would have been possible (but not useful—recall the principle of relevance!) to develop a question for each recommendation offered in this book. Instead, the questions presented are meant to identify how well a program produces the most frequently used display formats, and to isolate aspects and features that may be of particular importance to you. The checklist is designed to help you compare the relative virtues and drawbacks of a number of programs; it is not meant to make up your mind for you, but to focus your attention on what's important to you in a graphics package.

To use the checklist, score each *yes* answer with 1 point if the feature in question is one you would use sometimes, with 2 points if that feature is very important to you. Similarly, subtract 1 or 2 points as appropriate for each *no* answer. If the feature is of no concern to you, no points need be added or subtracted; alternatively, if the presence of a feature is a strong selling point with you, you may wish to award it more than 2 points. Comparison of the total scores awarded to different programs will tell you how useful a given package is for your purposes.

The checklist is in four parts. The first focuses on how easy it is to use a program, without considering the quality of the displays it produces. The second, third, and fourth parts deal with specific aspects of a program that make the product more or less effective. These last three parts of the checklist will bring to light most of the important features of a package.

Ease of Use

Talk to someone who uses the program you are evaluating. Many of these questions are best answered by the experience of someone who has passed through the initial stages of learning a program and has become adept at its use.

1. Does the program produce the display formats you use?

2. Does the program allow the generation of novel types of displays?

3. Can usable graphs be produced within 90 minutes of opening the package?

4. Does the display on the screen look the same as what is printed out? The WYSIWYG ("what you see is what you get") philosophy is popular in the Macintosh world and not popular in the UNIX world; programs for PC machines vary widely. This philosophy is strongly recommended for display design because you will sometimes have to make several successive drafts; if this can be done on the screen, so much the better.

5. Do you see the graph as it is being set up, so that you can alter its design as soon as a flaw appears?

6. Does the program require less than an average of 5 minutes to construct a simple line graph?

7. Does the program save most of the information used to plot a graph so you do not have to reenter it in order to plot data in a new way?

8. Is the program integrated with an effective, useful, spreadsheet program?

9. Can data be entered in more than one way? After you are familiar with the program, you probably will want to enter data in a batch mode (as a separate file, rather than by retyping it a part at a time).

10. Does the program have many ways of entering commands? This sort of flexibility should not be underestimated. Although interactive modes are helpful in some circumstances, especially when you are first learning, using them is often slower than specifying lots of information in a single batch at the outset.

11. Does the program make it easy to plot subsets of data?

12. Can you easily alter which independent variable is on the X axis and which is the parameter?

13. Can you easily alter what type of content—bars, lines, or points—is used?

14. Does the program have a useful Help function? A useful Help function is one that can be accessed online (without disrupting what you are doing); is organized so that one can easily find the information needed; and does not assume that you know too much. Is the program so well designed that you do not need to use the Help function?

15. Is the program so well designed that you do not need to read more than a few pages of instructions?

16. Is the program written by a reputable software house, which is likely to support it over the long-term and release periodic upgrades?

Creating the Framework

Graphs in L or T frameworks

1. Does the program automatically scale the range of values along the Y axis?
2. Can you override a function that automatically scales the range of values along the Y axis?
3. Can you control the origin of the Y axis?
4. Are tick marks placed at regular intervals (e.g., multiples of 10), instead of at locations of plotted values?
5. Are tick mark labels always round numbers?
6. Is every fifth or tenth tick mark drawn more heavily?
7. Can an inner grid be added or omitted?
8. Is the inner grid drawn with finer lines than used for the outer framework?
9. Is it possible to add a right Y axis to an L-shaped framework?

Pie graphs

1. Can you control which slice or slices are exploded?
2. Can pies can be drawn using two dimensions, or must they be drawn in three?
3. Is genuine projective 3-D used, not "pseudo 3-D" (which simply portrays parts as seen from different points of view)?

Creating the Content

1. Can you control the hue, saturation, and intensity of colors used for content elements?
2. Does the program automatically use discriminable colors for different content elements?
3. Does the program automatically vary intensities of colors?
4. Can the program automatically use colors that have about the same salience?

Bar graphs

1. Are bars automatically drawn with clearly discriminable internal hatching?
2. When three-dimensional bars are drawn, are they placed in a three-dimensional framework?
3. Can you adjust the spacing of bars?
4. Can you adjust the width of the bars?
5. Can you easily change the order of the bars?

Line graphs, layer graphs, and scatter plots

1. Are the leftmost and rightmost dots always plotted within the framework?
2. Are lines automatically drawn in clearly different patterns?
3. Are the dots in line graphs or scatter plots automatically drawn using clearly different symbols?
4. Can you control which symbols are used to plot lines and dots?

Step, pie, stacked-bar, and divided-bar graphs

1. Are different regions (wedges, slices, steps, etc.) automatically drawn so that they are visually distinct?
2. Can you control how wedges are filled with shading or hatching ?
3. Can you control how segments are ordered?

Creating the Labels

A good program will place labels correctly, as described in Chapter 3; it will also allow the user to adjust labels to prevent clutter, ensure detectability, and so on.

1. Does the program allow you to move labels?
2. Can the size and other typographic properties of labels be adjusted?
3. Is the title automatically drawn using letters that are typographically distinct from the other labels (but not all uppercase letters)?
4. Are the smallest labels clearly readable when printed out?
5. Are discriminable colors used to associate labels with content elements?
6. Can you control the salience of colors used to group labels and content elements?

Graphs in L or T frameworks

1. Are axis labels centered next to the relevant axis?
2. Can you add spaces where you want them between labels and content elements?
3. Are even long labels clearly associated with the correct tick mark on the X axis?
4. Are labels placed closest to the correct content element, and do they all appear in the same region of the display?
5. Can you choose when to use a key?
6. Are labels automatically placed so that they do not cut across an axis or content element?

Pie graphs

1. Are labels automatically placed consistently for each wedge, or are some inside and some outside the wedges?
2. Can you control the locations of labels?
3. Can you control whether or not a key is used?

Creating Multiple Panels

1. Can you choose which data to plot in multiple panels?
2. Can you easily arrange multiple panels on a page?
3. Are corresponding segments automatically ordered the same way in multiple displays?
4. Can you adjust the scales of the individual panels?

appendix 3
Summary of Principles and Their Psychological Bases

The three maxims discussed in Chapter 1 are: *The mind is not a camera; the mind judges a book by its cover; the spirit is willing, but the mind is weak.* This appendix summarizes the principles that are the offspring of these maxims. The interested reader is referred to Kosslyn and Koenig (1992) for a fuller discussion of the recent discoveries about the brain mechanisms underlying vision and memory.

The mind is not a camera

Our visual systems are not like cameras, which record what they are pointed at in a relatively veridical way. We actively organize and interpret what we see, in ways described by the following principles.

Salience. Large changes in visual qualities such as size or darkness attract attention. Neurons are difference detectors, and specific parts of the visual system draw attention to regions where there are changes in the stimulus (see page 24).

Levels of acuity. We cannot help but pay attention to patterns that are registered by the same "input channel." The visual system registers variations for lines at a given orientation at multiple levels of scale; each "channel" is sensitive to changes in regular light/dark alterations within about a 4-to-1 ratio (see pages 3–5, 140–141, 172–174).

Orientation sensitivity. The visual system immediately registers differences in orientation of at least 30 degrees. Neurons at various levels of the visual system are tuned to respond only to edges or shapes at particular orientations; in the initial stages of processing, the neurons have rather broad "tuning curves" and respond to edges within a relatively wide range of orientations. Thirty degrees is large enough to ensure that different input neurons will be used to encode patterns of different orientations (see pages 172–174).

Perceptual organization. The visual system automatically organizes input into psychological units. Marks that are nearby (proximity), arranged in a continuous pathway (good continuation), alike in shape or color (similarity), moving in the same way or direction (common fate), or arranged in regular patterns (law of good form) are organized as single perceptual units. The details of the mechanisms underlying this process are not known, but they presumably rely on the same properties of neurons that cause visual beats—namely, the neurons that register the same property form a unit, and neurons that register different properties specify boundaries (see pages 6–7).

Integral dimensions. Variations along some perceptual dimensions cannot be registered without our also noticing variations on other dimensions. In particular, height and width are attended to together, as are hue and saturation and hue and intensity. Presumably integral dimensions are processed by the same mechanisms and so cannot be dissociated (see page 116).

Perceptual distortion. Some visual dimensions are systematically distorted, notably area (which is progressively underestimated as area increases). Intensity and volume are also systematically underestimated. In contrast, line length is registered relatively accurately, although vertical lines appear longer than horizontal ones of the same length. These properties appear to spring from the way neurons code inputs, although the details of this process are not yet clear (see pages 5–6, 24, 114–115, 164–165).

Imprecision. Shapes and locations are not precisely conjoined during perception. Imprecision in judging spatial relations apparently occurs because different brain systems are used to register object properties (such as shape, color, and texture) and spatial properties (such as location, size, and orientation; see page 20).

Three-dimensional interpretation. If at all possible, the brain depicts a pattern as representing a three-dimensional object. The brain has built into it certain "assumptions" about the world, and we will see specific properties if these assumptions are met (Marr, 1982). If lines are drawn that mimic the edges we would see if an object were visible, we will see depth even when the edges are on a two-dimensional page. This is why the third dimension can be depicted on a two-dimensional surface—but the process is not perfect. Three dimensions can be depicted well enough on a flat surface to convey general impressions of differences and trends, but not high-resolution data (see pages 176–177).

The mind judges a book by its cover

The injunction *not* to judge a book by its cover is an attempt to fight our natural tendency to do just that; we take appearance as a clue to the reality. The principles of compatibility, cultural convention, and informative changes are refinements of this observation.

Compatibility. The Stroop experiment showed dramatically that if the physical appearance and interpretation (in this case, the color of ink and the meaning of the word written in it) conflict, the subject is impaired.

The brain attempts to fit all the information presented to it into a single coherent framework and expends effort to do so (see pages 8, 36–37; see Kosslyn & Koenig, 1990, for a discussion of this "constraint satisfaction" process). Therefore, the appearance of a pattern should be compatible with what it symbolizes.

Cultural convention. Displays are interpreted using common associations of a culture. This principle is in many ways similar to the principle of compatibility, except that the focus is on learned associations, not endemic properties of input. To identify an object is to access stored associations in memory; these associations are activated "automatically" and thereafter affect processing (see page 38).

Informative changes. Any change in a display is interpreted as conveying information. This principle follows directly from the basic "difference detector" nature of the mechanism: The brain registers differences and tries to interpret them (see pages 26–27).

The spirit is willing, but the mind is weak

Human visual and memory systems have certain limitations, which must be respected if a display is to be interpreted correctly. The following principles characterize these limitations.

Detectability. Marks must be large enough or heavy enough to be noticed. Nearby neurons that detect edges inhibit each other. If a mark is not large or heavy enough, the cells that typically would detect it are inhibited from responding (see pages 64–65, 88–89).

Discriminability. Two visual properties must differ by a large enough proportion or they will not be discriminated. Brain cells detect differences, and differences are signaled by changes in their activity. These neurons are not perfect, and so their rates vary due to chance, producing "noise" in the system. A difference must be large enough to cause a change in neural activity that cannot be confused with noise (see pages 90–91).

Limited short-term memory capacity. We can hold in mind only about four perceptual units, or other sorts of chunks, at the same time. Short-term memory is a dynamic state; information is retained because neurons are actively firing. By analogy: If one is juggling, the number of balls one can keep in the air at once depends on how high one can throw the balls, how quickly one moves one's hands, and on how quickly the balls fall. Neurons quickly adapt (fall, in the metaphor), and continue to fire only if repeatedly stimulated (thrown up), but there are limitations on how quickly they can be restimulated (how quickly the hands move). These limitations may involve the speed with which certain chemicals are produced and absorbed (see pages 8–9, 54–55; Baddeley, 1986).

Limited processing capacity. The brain has a finite capacity to process information, and if a task is too demanding we not only take more time, we also make more errors. Neurons in the parts of the brain that subserve vision, memory, and reasoning not only require oxygen and glucose to operate (quantities of both of which are limited) but also must be stimulated by other parts of the brain. There are certain "arousal" systems in the brain, which regulate alertness and mental energy. These systems can stimulate only a certain number of brain regions at the same time, because of limitations in the availability of necessary chemicals (see pages 28–29).

Relevance. The reader expects neither more nor less information than is necessary to answer a specific question. The person making a query enters a kind of social contract with the person being addressed, part of which is the assumption that the response will relate to the question (Grice, 1975). We are not passive viewers, but rather build up expectations about the world as we see it. These expectations lead us to seek specific information (Gregory, 1970; Neisser, 1967, 1976), and if the information we seek is not initially available, we will keep looking (for a while)—a task that requires effort (see page 21).

Appropriate knowledge. A reader can interpret a display only if the appropriate information is stored in memory. This information is used to identify the components of a display and to know how to reason about the relations among those components. The very concept of "identification" requires that the appropriate knowledge be stored, and all learned strategies for reasoning also correspond to stored information (see page 22).

notes

how to use this book

p. ix **several alternative displays** Cleveland (1985) elevates the idea that graph production is an iterative process to the level of a principle; he suggests regraphing data repeatedly and deciding on the final design after viewing the alternatives. This is reasonable advice if you have large amounts of data that can be organized in various ways, or if you want to explore the utility of alternative display formats that are appropriate for that type of data.

chapter 1

p. 2 **ineffective graphs** Johnson, Rice, & Roemmich (1980) reviewed the graphs in annual reports of fifty Fortune 500 companies and found that almost half the reports (21) contained at least one graph that was inappropriately constructed. However, the fact that the most common errors distorted recent trends suggests that not all errors could be attributed to simple ignorance of how to make good displays.

p. 2 **deficiency of how-to books** One notable exception is the excellent book written by Cleveland (1985). Cleveland's orientation, similar to mine, differs in the following ways: He does not adopt a step-by-step approach, and he offers fewer specific recommendations, discusses a wide range of exotic and special-purpose displays, often offers mathematical treatments of the material, and often focuses on uses of graphs in data analysis. Cleveland is a statistician and has a statistician's sensitivity to the nuances of data; I highly recommend this book, and the book by Chambers, Cleveland, Kleiner, & Tukey (1983), to anyone interested in using graphics to analyze data. However, the goals of these books are slightly different from that of my book; this difference is brought to the fore by a remark made by another statistician (Tukey, 1977, p. vi), who commented that a good pictorial display of data *"forces* us to notice **what we never expected to see"** (Tukey's italics and bold). For my book, I would say that a good graph forces the reader to see the information the designer wanted to convey. This is the difference between graphics for data analysis (Cleveland, Tukey, and others) and graphics for communication (my book). Given the present goal, I avoid novel or exotic types of displays, and focus on the types that are used most commonly in business and communication.

p. 2 **visual appeal** For many examples of visually interesting graphs, as well as some guidelines for designing them, see Bertin (1983), Holmes (1984), and Tufte (1983, 1990); for a review of five books on graph design, see Kosslyn (1985).

p. 3 **"classic" findings** For overviews of the psychology of perception, see Bloomer (1990), Dodwell (1975), Frisby (1980), Gregory (1966, 1970), Hochberg (1964), Kaufman (1974), Osherson, Kosslyn, & Hollerbach (1990), and Spoehr & Lehmkuhle (1982). For overviews of cognitive processes, see Anderson (1990), Glass & Holyoak (1986), Osherson & Smith (1990), and Posner (1978).

p. 5 **defocusing and clarity of image** For more information on this phenomenon, see De Valois & De Valois (1988); Julez (1980).

p. 5 **"false stereo" effect**   See Held (1980), Travis (1991), and Vos (1960). I discuss this effect in more detail in Chapter 7.

p. 6 **grouping effects**   The details of the mechanisms underlying these grouping processes are not known, but they presumably rely on excitatory and inhibitory connections that join neurons in the same state (e.g., responding to red) into a unit. These "coalitions" of neurons would also distinguish a unit from those formed by joining neurons with different values.

Tufte (1983) recommends eliminating all ink that does not convey information—even if this results in the elimination of symmetry, closure, and other simplifying properties. Tufte calls extra ink "chartjunk" and computes a "data-ink" ratio: the amount of ink used to convey data compared to total ink used in the display. Graphs that have a low data-ink ratio he terms "boutique" graphics. Spence (1990) points out that, contrary to this advice, more ink may allow people to read displays more quickly in some circumstances. For example, in one of his experiments Spence found that subjects could compare two boxes faster than two vertical lines. As a general rule, additional ink should be helpful if it completes a form, resulting in fewer perceptual units. Moreover, Carswell (1992), in her review of the literature on graph reading, found little support for the importance of a high data-ink ratio. In my opinion there is a place for visually interesting graphs, even if they contain more than the bare essential amount of ink, provided that the decorative material does not interfere with the information-conveying material.

p. 7 **embedded geometric figures**   See Reed & Johnsen (1975).

p. 8 **Stroop phenomenon**   For an excellent review and further discussion, see MacLeod (1991).

p. 12 **"Nutritional information" graph**   Kosslyn (1989) found 26 specific problems with this display, evaluating it with respect to the principles and recommendations discussed in this book.

p. 12 **psychological principles and visual displays**   A word on my use of the term "principle": Properly speaking, a principle ought to specify a key feature of the mechanisms that produce a phenomenon. In some cases, however, the principles I summarize are in fact descriptions of regularities in data, which presumably reflect specific features of the underlying mechanisms. This book is intended as a practical guide to graph design, and I have avoided getting into details of mechanisms if they would not further this end. For more detailed treatment of the principles offered in this book, see the Bibliography.

p. 13 **flawed displays in studies**   For a particularly thoughtful review, see Macdonald-Ross (1977).

p. 13 **contradictory studies**   For reviews of this literature, see Carswell (1992), DeSanctis (1984), Feinberg (1979), Jarvenpaa & Dickson (1988), Macdonald-Ross (1977), and Wainer & Thissen (1981).

p. 13 **essential message**   This book guides the user to develop effective graphs for normal adults. There is a literature on graph effectiveness in children, which indicates that the quantitative specifications of some of the principles discussed here should be modified for them; see, e.g., Bryant & Somerville (1986), Curcio (1987), and Mokros & Tinker (1987). There is also a small literature on graphs for the blind. Such graphs appear to be more effective if they are in concordance with the corresponding tactile principles. See, e.g., Aldrich & Parkin (1987); Lederman & Campbell (1982, 1983).

p. 13   **generality of principles and recommendations**   Although all the recommendations I offer here are consistent with the available literature and follow from the principles I develop, in some cases the connections are indirect. Clearly, this book is not the last word on how visual displays should be designed. I hope that this will be the first of many editions, each successive one taking advantage of additional research results to develop ever more rigorous guidelines for display design.

p. 14   **the same components** The idea that graphs can be broken into components is discussed from a slightly different perspective by Winn (1987) and in more technical terms by Kosslyn (1989).

p. 14   **definition of "content"** Kosslyn (1989) uses the term "specifier" (also adopted by Carswell, 1992) instead of "content" because in a sense the entire display is the content. However, to avoid jargon, I will use the term "content" in a more restricted sense here.

p. 15   **parameter** The term "stratum" is sometimes used for "parameter" (see, e.g., Lewandowsky & Spence, 1989).

chapter 2

p. 21   **see both clearly** In addition, we may have trouble using precise spatial information in recognition because such information typically is used to guide actions (such as reaching for the handle of a coffee cup, moving the eyes to notice who just came in a door). This sort of information is rarely used to recognize objects, and it requires effort to use it in this way (see Kosslyn & Koenig, 1992, chapter 3).

p. 21   **use a table** There are a surprisingly large number of studies that were designed to compare the relative efficacies of graphs and tables of numbers (for reviews, see Casali & Gaylin, 1988; DeSanctis, 1984; Jarvenpaa & Dickson, 1988; MacGregor & Slovic, 1986). The general finding is that graphs are better than tables of numbers only for specific purposes. For example, Washburn (1927) found that tables of numbers are better for recall of specific amounts, whereas pictograms are better for simple comparisons, bar graphs for complex comparisons, and line graphs for trends (as cited in Umanath & Scamell, 1988). Umanath & Scamell (1988) compared a table to a bar graph and found that the graph was better for recall of rank order and pattern information, whereas the table was as good as the graph for specific point values; similarly, Spence & Lewandowsky (1991) found that both pie and bar graphs were superior to a table when relative values had to be compared—even if only a few numbers were involved (counter to the suggestion of Tufte, 1983, to use a table for small amounts of data). However, although Benbasat & Schroeder (1977, cited in Lucas, 1981, p. 758) found that graphics displays were better than tables in an inventory management task, others (e.g., Lucas, 1981) found no clear difference. And yet Moriarty (1979), Stock & Watson (1984), Nawrocki (1973), and Schwartz (1984) all found that subjects performed better with graphics displays than tables in decision-making contexts. In addition, Casner & Larkin (1989) showed that a graphical format was superior to a table when subjects had to use data to make airline reservations. The graphs they used were designed to reduce the effort of using specific mental processes and to reduce the amount of search necessary; these considerations have been incorporated into the principles and recommendations offered in this book. In short, although graphs are not always superior to tables, the bottom line—as I read it—is that graphs are more effective than tables when relations among values are critical (cf. Jarvenpaa & Dickson, 1988; Winn, 1987). Note, however, that the effectiveness of tables—like graphs—depends on how they are designed (see Ehrenberg, 1975).

p. 21 **rules of discourse** Grice (1975).

p. 21 **principle of relevance** This principle presumably arises out of an implicit "social contract" that governs discourse; it is one of the rules for playing the "language game."

p. 22 **identifying patterns** Our brains did not evolve for the purpose of reading graphs, but mechanisms that evolved for other purposes can be used to this end. For example, consider what it means to identify an apple. If you see an object and then identify it as an apple, that means you now know more about it than is immediately apparent (that it can be eaten, that it grows on trees, that it has seeds inside, and so on). How can you know these things, given that all you see is a red spheroid of a certain size, texture, and reflectance? The shape, color, texture, and size all match information stored in memory about a specific previously seen object. Associated with this object are a variety of facts, which are available once you match the input to the appropriate stored information. Thus, to identify the apple, the input must match the appropriate information in memory, which in turn activates other information that is associated with such objects. The same process presumably is used to identify familiar patterns in visual displays of quantitative information. As expected, given the necessity of prior experience, there is ample evidence that the effectiveness of graphic displays varies for different types of students and different materials (e.g., for a review, see Winn, 1987).

p. 22 **significance of the pattern** Pinker (1990, p. 109) identifies 12 visual patterns of lines that signal distinct quantitative trends: flat (unchanging), steep (increasing rapidly), inverted U (quadratic), U (quadratic), jagged (random), undulating (fluctuating regularly), straight (linear), s-shaped (cubic), rectilinear (abruptly changing), not flat (variable X affects variable Y), parallel (variable Z has additive effects on those of variable X), converging (variables X and Z interact). These patterns can be more finely distinguished; for example, there are four distinct ways in which lines can converge. In short, an expert graph reader has a remarkable amount of knowledge about interpretation of patterns of lines.

Pinker argues that graphs are useful in large part because experienced readers develop specific "graph schemata" in memory, which allow them to identify meaningful patterns. Simkin & Hastie (1987) develop this idea and present empirical support for it (see also Wainer & Francolini, 1980). Winn (1983), in his studies of eye-movement patterns, found that people quickly fixate on "meaningful" shapes when information is presented graphically. In addition, it is worth noting that Phillips, Coe, Kono, Knapp, Barrett, Wiseman, & Eveleigh (1990) found that the precise nature of symbols matter more to inexperienced readers (of maps, in their study) than to experienced readers. Moriarity (1979), Stock & Watson (1984), and MacGregor & Slovic (1986) show that relatively exotic displays can be used effectively if readers learn how to interpret them. See also Ehrenberg (1975), Jarvenpaa

& Dickson (1988), Simkin & Hastie (1987), and Vernon (1946) ; but also see Macdonald-Ross, 1977, for a critique.

p. 23   **this aspect of understanding**   Even if the data do not produce easily identifiable patterns, familiar formats are easy to read because the reader can effortlessly carry out a sequence of steps. A demanding task entails many steps and requires one to attend to each of them. As one becomes familiar with a task, it becomes less demanding because one organizes the steps into larger units and no longer needs to attend closely to the individual steps; with practice, we develop "automatic" processes. For example, when first learning to drive the student must pay attention to every detail; shifting gears with a manual transmission requires pressing in the clutch, letting up on the gas, pulling the stick, pressing down on the gas, and slowly letting up the clutch. With practice, the entire process becomes organized as a single operation, and one no longer needs to pay attention to the details (see Anderson, 1981, 1990; Osherson et al., 1990). Indeed, these kinds of habits may even use a different brain circuit, which directly connects inputs to responses (see Mishkin & Appenzeller, 1987).

p. 23   **less well-known displays**   I discuss here only those types of graphs that are commonly used in newspapers, magazines, and other nonspecialized media; for a more sophisticated and detailed taxonomies and discussions of special-purpose displays, see Bertin (1983), Chambers, Cleveland, Kleiner, & Tukey (1983), Cleveland (1985), Fienberg (1979), Macdonald-Ross (1977), Tufte (1983, 1990), Tukey (1972, 1977), and Wainer & Thissen (1981). Keep in mind, however, that although some of these exotic displays may be better than conventional displays in laboratory tasks, they may fall short in more natural settings (for example, Goldsmith & Schvaneveldt, 1985, found a "star" display to be better than a bar display in the lab, but Peterson, Banks, & Gertman (1982, found no difference in the context of a nuclear power plant). My own view is that the conventional displays have survived a kind of Darwinian winnowing process, and the mere fact that they continue to be employed over so many years (since their invention by Playfair, 1786, in some cases) is itself evidence of their utility. For the interested reader, Cleveland (1987) provides an annotated bibliography of research and theory on statistical graphics. For a detailed discussion of why some displays are processed better for some purposes than others, see Simkin & Hastie (1987).

p. 23   **choice of graph format**   Gnanadesikan (1980; cited in Wainer & Thissen, 1981, p. 196) introduced a number of criteria to consider when choosing a display, such as its potential for internal comparisons and aid in focusing attention. Many of his criteria are similar to those offered here. However, Gnanadesikan appears to have neglected a central point I wish to stress: The usefulness of a graph can be evaluated only in the context of the type of data, the questions the designer wants the reader to answer, and the nature of the audience.

p. 23   **relations among components**   Cleveland & McGill (1984a, 1985, 1986, 1987) suggest that visual dimensions can be ordered in terms of how well people can use them to compare quantitative variations. Based on a mixture of theoretical and empirical findings, they suggest the following ordering of dimensions (from best to worst): position along a common aligned scale, position on identical but nonaligned scales, length, angle/slope, area, volume/density/color saturation (all about the same), hue. Carswell (1992) reviews the results of 39 experiments and reports some support for this ordering. However, she found minimal differences for position, length, or angle, but area and volume were consistently worse than the other dimensions; these findings suggest not so much an ordering as two categories, with the members of one—area and volume—being inferior to members of the other. In addition, Carswell found that the ordering of visual dimensions depends in large part on the type of task. This ordering fares better at predicting performance when subjects must focus on one portion of the graph (as when specific point values were extracted) than on tasks where different portions must be integrated or compared; indeed, the predictions of Cleveland & McGill's ordering were actually contradicted when subjects had to synthesize information from different portions of a display (e.g., when determining whether the variability of the data points was large). Also, the model did not fare well when subjects had to recall graphed information.

Spence & Lewandowsky (1991) report findings that suggest that Cleveland & McGill's ordering reflects very low-level "preattentive processing," which affects only the initial phases of registering a display (see also Cleveland, 1985; Cleveland & McGill, 1987). Spence & Lewandowsky asked subjects to use different types of displays to make local and global comparisons within a specified period of time (under a "deadline"); the model predicted the results best when the deadline was very short. In addition, Simkin & Hastie (1987) also report that the relative efficacy of different visual dimensions for conveying information about quantity depends in part on the specific task. Moreover, Petersen & Schramm (1954) asked aviators to use graphs that specified information with angles or position on aligned scales, and found that their performance tended to be better with angles. These findings are interesting because other types of subjects show the opposite pattern. Carswell (1992) argues that because aviators are trained to read circular displays, these results may provide evidence that training affects the relative ease of using specific types of content. Indeed, DeSanctis & Jarvenpaa (1985) showed that with practice business students were better able to use graphs to make financial forecasts—but that practice only improved their performance with a "standard" horizontal bar graph format, and not one with scales that had different maximal amounts and nonround value labels. It is clear that the accuracy of simple judgments of amount is not the only factor you should consider when deciding how to display data. The recommendations offered in this book are based on the wider set of principles summarized in Appendix 3.

p. 24 **because it is difficult to measure angles** Cleveland (1985) relies on this observation when he recommends that pie graphs should not be used at all. However, his example (p. 264) relies on comparisons of wedges that differ on the order of about 3%, which probably are not significant differences—and hence the visual impression of no differences may accurately reflect the lack of systematic variation in the data. If subtle differences are significant, I recommend either using a bar graph (provided that the relation of parts to the whole is not critical) or a divided-bar graph with the values marked directly on the segments.

p. 24 **perceptual distortions** In addition to systematic distortions, our visual systems are also prey to illusions—we see properties that are not present (e.g., bent lines when straight ones exist). See Frisby (1980), Gregory (1966), and Robinson (1972). For discussion of the role of such illusions in graphs, see Graham (1937), Kolata (1984), Poulton (1985), Schiano & Tversky (1992), and Tversky & Schiano (1989).

p. 24   **visual impression of area**   There has long been a controversy about just how well human beings can determine the area of pie wedges. Eells (1926) reported the first study of this ability, finding that people read the relative proportions of pie wedges more accurately than segments of divided-bar graphs, particularly when more components are included (but see Von Huhn, 1927, for a critique). Croxton & Stryker (1927) report similar findings, but found that divided-bar graphs were better than pie graphs in some circumstances (when there were only two parts, and they did not display a 50:50 or 75:25 relation). Macdonald-Ross (1977) recommends bars over pie graphs because people systematically underestimate area. However, Macdonald-Ross's recommendation was based on judgments of the relative size of entire forms, not their components. Similarly, Cleveland & McGill (1984a) found that people cannot judge area well from pie graphs, but they asked subjects to judge the percentage a smaller element was of a larger one—which is not what we usually do when reading graphs. Simkin & Hastie (1987) found that subjects could compare two quantities (and determine what percent of the larger is represented by the smaller) more accurately for bars, then divided bars, and then pie segments; in contrast, when asked to determine the percentage of the whole, subjects judged segments of pies and bars more accurately than segments of a divided bar. (However, Simkin & Hastie did not include a scale on the divided-bar graphs, which probably would have facilitated performance for these displays.) Spence (1990) asked subjects to judge the relative proportions two parts are of a whole (rather than direct size comparisons) and did not find systematic underestimation when only one dimension (e.g., wedge size, height) was varied. Indeed, Spence found that subjects could compare the relative proportions of pairs of bars or pairs of pie elements equally well when only a single dimension was varied, and there was no clear association between the number of dimensions of a stimulus and the subjects' accuracy. However, Spence found that a pies were not quite as good as bars when viewers had to switch back and forth between different formats. But Spence & Lewandowsky (1991) found that a pie graph can actually have a slight advantage over a bar graph if the judgment requires complex comparisons of components. In addition, Croxton & Stein (1932) report data that show that subjects can compare the areas of two bars better than of two squares, circles, or cubes, but that the estimation of squares and circles is about the same.

The waters are muddied in part because different researchers adopted different tasks and used different materials. The recommendations I offer here are consistent with the major themes of this literature, but I have given more weight to the results from studies that used more realistic tasks and displays.

p. 24   **underestimation of area**   Similarly, when people estimate how bright a point is on a screen, they systematically report points as dimmer than they are—and this underestimation becomes more severe as stimuli become brighter. Thus, increasingly larger increments of area or brightness must be added to the starting level to get the same psychological (perceived) increase as the overall amount increases; the forms must be systematically larger or brighter than they would be if we were accurate perceivers. Such systematic distortions are captured by Stevens's power law, which states that the psychological impression is a function of the actual physical magnitude raised to an exponent (and multiplied by a scaling constant). To be precise, the perceived area is usually equal to the actual area raised to an exponent of about .8, times a scaling constant (see also Brinton, 1916;

Macdonald-Ross, 1977); the perceived brightness is usually equal to the actual brightness raised to an exponent of about .7, times a scaling constant. In contrast, relative line length is perceived almost perfectly, provided that the lines are oriented the same way (see Baird, 1970; Stevens, 1974, 1975). (Note, incidentally, that pain has the opposite relation to reality: Increasingly less electricity is required to induce an increment of discomfort as the general level increases.)

It is important to note that Teghtsoonian (1965) found that the exponent for area varied depending on exactly what the subjects were instructed to do: The exponent was close to 1 (representing veridical perception) if subjects were asked to judge "real," as opposed to "apparent," area. Spence (1990) found that exponents were close to 1 when subjects compared the relative proportions of two pie wedges, vertical lines, horizontal lines, disks (pies seen from above and to the front), bars, boxes and cylinders, provided that only one dimension was varied at a time. Spence did not explicitly ask subjects to judge the "real" extents, but suggests that the use of graphical elements alone may have led them to make judgments of real, not apparent, size. In addition, Meihoffer (1969, 1973) showed that the errors in estimating area are virtually eliminated if subjects are given a scale and areas vary only in the discrete (and easily discriminable) increments that are illustrated in the scale.

Although the distortion can be overcome, this may require motivated readers who are willing to reason their way past initial visual impressions. In general, such impressions will be distorted: The larger two regions are, the greater the percentage of difference will have to be in order to be seen as equivalent to a difference between two smaller regions. You cannot correct for this error by distorting the graph, because individuals differ in the degree of error and the degree of error depends on the precise material being compared (see also Cleveland, Harris, & McGill, 1982).

p. 24 **one-fourth of graph readers** Eells (1926) found that most (51%) of his subjects read relative areas of pie graphs by looking at the arcs, 25% by estimating area, 23% by estimating the angle at the center, and 1% by estimating the length of chords. Unfortunately, not only do our visual systems distort area, but they also fail to register angle with great precision. Relatively small acute angles tend to repulse each other (see Howard, 1982; see also Schiano & Tversky, 1992). In addition, angles that are symmetrical around the horizontal axis are seen as larger than angles of the same size that are symmetrical around the vertical axis (Maclean & Stacey, 1971). Divided-bar graphs are not prey to these problems, but Eells (1926) found that almost three times as many people preferred pies to divided bars. (These data are admittedly a bit dusty more than 65 years later, but there is no reason to think that people have changed in the interim; indeed, pie graphs are probably even more common today than in 1926.) So if you are more interested in conveying precise amounts than in displaying components of a whole, consider using a bar graph. Cleveland & McGill (1984a) argue that people can read grouped bar graphs (with a common baseline) more accurately than pie or divided bar graphs. But bar graphs do not provide an immediate impression of how parts form a whole, and so are not ideal if this is your goal.

p. 24 **power of a visual display** The reader no longer apprehends the differences at a glance, but must mentally add and subtract numbers. If relatively precise numbers are needed but you want the reader to see the amounts quickly, consider a divided-bar or bar graph.

p. 24   **respond to a change in stimulation**   For details, see Kosslyn & Koenig (1992), chapter 3.

p. 24   **salience**   For examples of the role of variations in salience in graphical perception, see Brown (1985) and DeSoete & DeCorte (1985).

p. 24   **informative changes**   For similar ideas, see Macdonald-Ross (1977); the classic work on the concept of "information" was written by Shannon (1948).

p. 26   **not passive viewers**   See Gregory (1970); Neisser (1967, 1976).

p. 28   **scale with divided-bar graph**   Cleveland & McGill (1984a) found that subjects do not estimate the lengths of segments of divided-bar graphs very well. Thus, it is critical that a scale be included if relatively precise impressions are required.

p. 28   **limited processing capacity**   See Kahneman (1973) for a discussion of the concept and a review of classic studies of such limitations.

p. 28   **properties of our brains**   When we think, our brains consume more glucose and oxygen, the parts of the brain that keep us alert have only a limited supply of crucial chemicals, and neurons can respond only at certain rates. The brain has only a limited capacity to process information, and if a task is too demanding we not only take more time but also make more errors.

p. 28   **read at first glance**   Simkin & Hastie (1987) offer a sophisticated theory of the specific processes that are used to read a graph. This theory specifies four "visual routines" that the mind "runs" on a representation of a graph in order to decode it. The routines allow the reader to find an "anchor point," to scan from that point, to project mentally a reference line (e.g., from one bar to another or from a content element to the Y axis), and to superimpose a mental image of one part of the content over another (e.g., pie wedges, to compare their relative sizes). This theory has yet to be tested in detail, but it seems clear that insights from theories such as this will produce another round of recommendations for graph design that take advantage of more detailed knowledge of what the reader does when decoding these displays.

p. 30   **isotype**   See Neurath (1974); Macdonald-Ross (1977) provides a nice discussion of the virtues of this format. I have not devoted much time to it, however, for practical reasons. Relative to the other types of displays, isotypes are difficult to produce. With continued progress in computer graphics packages, this may not long remain the case, however; if such displays become easy to produce, they often may be preferable to standard bar graphs.

p. 32   **choice between them**   Several researchers have explicitly compared performance when subjects were asked to extract point values or trends from bar graphs versus line graphs and sometimes report contradictory findings (for a review, see Jarvenpaa & Dickson, 1988; see also Casali & Gaylin, 1988). In some of these studies, the displays were not constructed well; the important question is whether different types of competently prepared displays are more or less effective for different purposes, and not much can be concluded when the displays are flawed (for discussion of this and related issues, see Macdonald-Ross, 1977). In addition, in some cases, subjects may be able to adopt special strategies to help them use a particular type of display, but at least some of these strategies may be laboratory-specific. The recommendations offered here are based on my critical reading of the literature and unpublished experiments conducted in my own laboratory.

p. 34   **expending time and effort**   Simcox (1983, summarized in Pinker, 1990, pp. 120–122) provided good evidence that bars are encoded in terms of their heights whereas lines are encoded in terms of their slopes. In the most straightforward experiment, subjects were asked merely to classify the heights of bars versus the heights of the ends of lines, or were asked to classify the slope or overall height of a line versus the slope or overall height of an implicit line connecting tops of bars. The subjects could encode the height at a point faster from bar graphs, and slopes and overall heights faster from lines. In addition, my advice about when to use line graphs is based in part on the findings of Schutz (1961a), who reported that people can determine the nature of a trend better if the data are presented in a line graph than in a bar graph. Provided the format does not mislead the reader, it makes sense to use the one that is easiest to read.

p. 34   **signatures of specific maladies**   The "handwriting" may change in the revised taxonomy of psychological disorders that is about to be released, which will require practitioners to learn the new categories. In so doing, they will also learn what is signaled by the corresponding visual patterns in the profile.

p. 36   **specific measurements**   For one demonstration that line graphs are harder to use than bar graphs when specific points had to be compared, see Culbertson & Powers (1959). We have found similar results in experiments in my laboratory. Tufte (1983) suggests that one should never use bar graphs because the bars contain redundant information (too much ink for the amount of data conveyed); he prefers a minimalist graph, illustrating only line segments that correspond to the tops of bars. I reject this opinion for three reasons: The content of such displays would not be very salient; the heights of line segments are not immediately translated visually into length (a visual dimension we register easily); and Carswell's (1992) review of the literature on experiments on graph perception did not support for the utility of Tufte's proposal.

p. 38   **labeling horizontal bar graphs**   Cleveland (1985) recommends a variant of a horizontal bar graph when each value is labeled. This type of graph has a dot at the location that would be the end of the bar, with a dotted line joining the dot to the Y axis. It is not clear whether such displays are to be preferred over the standard ones.

p. 38   **preference for vertical format**   In addition, Schutz (1961a) found that people could determine the nature of trends better when data were presented in a vertical bar-graph format than when they were presented in a horizontal bar-graph format; however, Spence (1990) found no compelling evidence that one format is better the other. It is clear that the precise task, the complexity of the display, and the nature of the audience all play important roles in determining which display is most easy read.

p. 46   **form a cloud**   Lewandowsky & Spence (1989) report that people required comparable amounts of time to read scatter plots with different numbers of points, and actually became more accurate when more points were included. Legge, Gu, & Luebker (1989) found that the ease of reading scatter plots depended only slightly on the number of points. It is clear, then, that readers tend to get an overall impression of the shape of the cloud, which in some circumstances is easier when more points are present. Legge et al. also found that subjects could estimate means and variances more easily from scatter plots than from tables of numbers or an alternative display in which luminance was varied to convey amount.

p. 46 **overwhelm a reader** This potential problem is especially severe for variants of the scatter plot that use special symbols to present a large amount of information in a single display (see, e.g., Cleveland & McGill, 1984b). One variant that appears to becoming more common is the "box-and-whisker" plot. In this type of graph, a box encloses the middle 50% of the data and "whiskers" ("I"-shaped error bars, of the sort to be discussed shortly in the text) extend to the extremes. A horizontal line in the box indicates the median (see also the "range charts" of Schmid, 1954). In addition, some scatter plots include symbols that vary in size to indicate the number of data points; this variation is intended to allow the reader to see which trends are likely to be most stable (see Bickel, Hammel, & O'Connell, 1975; see also Wainer & Thissen, 1981). However, such augmented scatter plots are often very complex visually, and many readers may not take the time to decode them. Consider your audience carefully before preparing such complex displays.

p. 54 **relatively small capacity** For more on the bases of memory limitations, see Baddeley (1986), Kosslyn & Koenig (1992), chapter 8, and Newell (1990).

p. 54 **four chunks** Miller (1956) originally suggested that we could hold seven chunks in mind, but later work has shown that the number is more like four (e.g., Ericsson, Chase, & Faloon, 1980).

p. 54 **in a separate display** This recommendation is based in part on results reported by Schutz (1961b), who found that subjects could compare lines (determine the one with the highest value at a particular point on the X axis) more easily when the lines were presented in a single display than when they were presented in separate displays. However, there was no difference when subjects extracted point values from only a single line. The advantage of comparing lines in a single display may have been due to the difficulty of remembering the lines in separate displays. At some point, probably when more than four perceptual units are present, this factor will be overwhelmed by the sheer confusion of too many lines in the display (see also Jarvenpaa & Dickson, 1988).

p. 56 **don't use a key** Milroy & Poulton (1978) compared direct labels on lines with keys that were placed in the lower right of the display or under the display. They found that readers could use direct labels quickest, without loss of accuracy, and they attribute this advantage to fewer processing steps and a lighter load on short-term memory.

p. 58 **measures of variability** Information about variability allows the reader to see how stable the means are (that is, to know the range in which the means would fall if the data were collected again in an independent sample), and to evaluate the depicted differences and trends accordingly. If the number of observations that went into each plotted data point is included (possibly in the caption), readers who know about "*t* tests" can determine whether differences are "statistically significant" (see, e.g., Games & Klare, 1967).

p. 58 **grid lines help the eye to focus on vertical extent** This effect was pointed out by Cleveland (1985); also see Cleveland & McGill (1985).

chapter 3

p. 66   **curious visual illusion**   See Poulton (1985) and also Hotopf (1966) for a discussion of the relation between this illusion and other types of illusions (see also Hotopf & Hibberd, 1989). In addition, it should be noted that our visual systems tend to distort the slopes of rising content lines so that they are closer to a 45 degree diagonal; we overestimate slopes of relatively shallow lines and underestimate slopes of relatively steep lines (Tversky & Schiano, 1989; Schiano & Tversky, 1992). This phenomenon is all the more reason to include an inner grid if you want readers to have an accurate impression of changes.

p. 66   **need for a second Y axis**   I recommend using inner grids, rather than duplicate Y axes, when the reader is to see relatively precise values because I assume that it is easier to trace along a grid line to a single Y axis than visually to track a shorter distance through empty space to a second Y axis. However, an experiment should be conducted to determine whether this is in fact the case.

p. 76   **put steps or bars next to each other**   Spence & Lewandowsky (1991) found that adjacency helped subjects compare bars but did not matter very much when they compared wedges of a pie graph.

p. 78   **leaving out a portion of the scale**   Cleveland (1985) worries that bars are misleading if the initial value is not 0 (and hence instead suggests using dotted lines that lead to a filled circle marking the total amount). In my view, this is not a problem: The important consideration is that the visual impression of differences and trends is compatible with the actual differences in trends in the data. Provided that this is true, and slash marks are used to indicate discontinuities in a scale, there is no reason not to use bars when part of the Y axis has been excised.

p. 80   **logarithmic scale**   For a more detailed discussion of logarithms and their use in graphs, see Cleveland (1985). Cleveland & McGill (1985) recommend using base 2 logarithms if fractions of exponents would be necessary with base 10; they also point out that base 2 makes increasingly good sense as more people use computers and develop intuitions about powers of 2. But the norm currently is still base 10.

p. 84   **vertical reference lines**   See Cleveland (1985) for further discussion of the utility of such reference lines.

p. 88   **minutes of arc**   A minute of visual arc is 1/60th of a degree.

p. 88   **read virtually perfectly**   Smith (1979) reports a large-scale study of legibility of letters. The smallest visual angle at which subjects could read letters was .0005 radians, the largest was .0127 radians; the mean was .0019, and the median was .0017 radians. The study was conducted under naturalistic conditions; the stimuli ranged from individual letters to words and text, and varied in typeface as well as size. Smith reports that 98% of letters can, in general, be read if the minimal size is .0046 radians, and that people read virtually perfectly when the letters appear at .007 radians. A radian is equal to the angle at the center of a circle when two radii are separated by a distance along the circumference equal in length to one radius (which is equivalent to approximately 57.295 degrees).

p. 90   **Weber's law**   See Baird (1970).

p. 90   **serif versus sans serif**   I have sometimes heard or read (e.g., on p. 13 of the style guide, *The basic elements of design*, that came with my Apple Personal LaserWriter) that titles and labels should be in typefaces without serifs (the short tabs and edges on the ends of some letters, such as d, f, and h). Fonts with serifs are said to be more complex than those that do not have serifs, such as Helvetica, which is shown on the text page. Serifs are supposed to help you read words more quickly by providing additional cues, but some believe that they actually make letters less discriminable. However, there is no good evidence that serifs consistently aid or impair reading under normal reading conditions. Indeed, Gould, Alfaro, Finn, Haupt, & Minuto (1987), Smedshammer et al. (1990, reviewed in Frenckner, 1990), and Zachrisson (1965) found no differences in reading speed for the two kinds of typefaces. For reviews of studies of different typographic fonts, see Patterson & Tinker (1940), Tinker (1963), and Zachrisson (1965). However, Tinker (1963) and others have found that people read italic print more slowly than nonitalic lowercase, and all uppercase letters were found to be read 10–15 percent more slowly than all lowercase letters.

p. 94   **unmarked terms for a dimension**   See Clark & Clark (1977). Note that other factors also affect word order, such as the "freezing principle" discussed by Cooper & Ross (1975) and Pinker & Birdsong (1979); according to this principle, the longer, more stressed term in a pair typically is second. In many—but not all—cases the shorter term labels the dimension, so the first word is the neutral one.

p. 96   **value labels on the Y axis**   There is no good reason to place value labels so that they are parallel to the Y axis. Coffey (1961) reports that people can read letters and numbers as easily when they are arranged in a column as when they are arranged in a row. Rotating the numbers will slow readers down (see Jolicoeur, 1990), which is not a good idea unless it is necessary.

p. 100   **label content elements directly**   In addition to the results of Milroy & Poulton (1978) summarized earlier, this recommendation is based on results reported by Parkin (1983), summarized in Pinker (1990). Parkin investigated the ease of answering questions about the relative heights and slopes of different lines at specific points on the X axis when five different methods of labeling were used: The label was next to the line somewhere along its length (and so was grouped with the line via the principle of proximity); was aligned with its end (and so was grouped with the line both via proximity and via good continuation); was aligned with its end but separated by a white space (and so was grouped with the line via good continuation but not proximity); was in a key; or was in a caption. Parkin also varied whether the labels were printed in the same or different color as the line. He found that the subjects were fastest when both good continuation and proximity were used to group the labels. However, this result depended on whether the lines formed relatively simple patterns or whether they crossed over many times. When the lines were clearly distinct, putting labels directly next to them was best; when lines formed complex patterns, the three methods that took advantage of grouping principles were equally good. It is of interest that direct labeling was clearly better than use of a key or caption for the simple patterns, but was much less superior for complex (and cluttered) graphs. In general, using the same color for lines and labels also facilitated performance.

chapter 4

p. 106 **facilitating comparison of wedges** If people typically read pie graphs by forming chords, it would be important to have wedges that are to be compared at similar orientations: The visual system distorts apparent line length when lines are seen at different orientations. However, as previously noted, Eells (1926) found that only 1% of his subjects read pie graphs by estimating the length of chords (51% looked at the arcs, 25% estimated area, and 23% estimated the angle at the center).

p. 112 **visual impression of a difference** Brinton (1919). See Jenks & Knos (1961) for a study of how greater quantities can be represented by increased amounts of ink; they found that the psychological impression of the amount does not increase in equal steps as equal amounts of ink are added. This is another example of the principle of perceptual distortion.

p. 114 **pictures of different heights** Macdonald-Ross (1977) recommends against varying the size of a picture to convey information; instead, he recommends varying the number of pictures (each one standing for the same fixed amount) to produce a row or column, the length of which specifies the total value. His reservations about varying size per se appear to be circumvented if the technique is used only to convey a general impression of differences and the display respects the principle of compatibility (small objects are not drawn inappropriately large) and is properly titled (or is accompanied by an explanatory caption).

p. 114 **overestimation of vertical bars** Graham (1937). Because of the distortion of vertical length, Jarvenpaa & Dickson (1988) advise using horizontal bar graphs if individual data points are compared. However, there is no evidence that the distortion of vertical lines seriously affects comparisons among vertical lines (all vertical lines are distorted the same way). For additional discussion of illusions, see Frisby (1980), Gregory (1966, 1970), and Robinson (1972).

p. 116 **integral dimensions** Garner (1970, 1974, 1976) suggests two criteria for determining whether two dimensions are integral: filtering interference (difficulty in ignoring one dimension while attending to the other) and redundancy gains (facilitation in reading one dimension when the value on the other provides redundant information). A weaker form of this relation occurs with "configural dimensions," which exhibit only filtering interference. For example, the angle formed by two lines is configural with the slopes of the lines (see Beringer, 1987; Carswell & Wickens, 1987a, b, 1988, 1990; Clement, 1978; Jacob, Egeth, & Bevan, 1976; Pomerantz, 1981).

p. 116 **covarying height and width** Carswell & Wickens (1990) discuss ways in which one might try to pack more information into a display using integral dimensions. However, these displays are useful only in limited circumstances, when the reader must attend to—and integrate across—several dimensions to make a decision (see also Barnett & Wickens, 1988; Carswell & Wickens, 1987a, b). Indeed, Casey & Wickens (1986) and Jones & Wickens (1986)—both cited in Carswell & Wickens (1990)—found that integral displays produced no better performance, and sometimes worse performance, than nonintegral displays even when the subjects had to process correlated variables (see also Coury, Boulette, & Smith, 1989; Sanderson, Flach, Buttigieg, & Casey, 1989). Moreover, it seems clear that the efficacy of integral or configural displays depends critically on how variables are assigned to specific dimensions or features of the display (MacGregor & Slovic, 1986).

An oft-cited use of configural information in displays was described by Chernoff (1973; see also Chernoff & Rizvi, 1975). Chernoff designed schematic faces in which the shape of each feature conveyed information about a separate variable; the overall expression of a face was intended to convey an impression about the relations among the variables. Jacob, Egeth, & Bevan (1976) and Wainer (1979a, cited in Wainer & Thissen, 1981) showed that people can evaluate the psychological "distance" between the faces. However, these displays are tricky—the features do not appear to be equally important in creating the overall impression (cf. MacGregor & Slovic, 1986), and sometimes may be difficult to read. Numerous variants have been proposed (e.g., Wakimoto, 1977; Wainer, 1979b).

In short, although using integral and configural dimensions is sometimes an effective way to pack much information into a single display (e.g., Bickel, Hammel, & O'Connell, 1975), readers typically need considerable practice before being able to decode such displays easily.

chapter 5

p. 119   **bar graphs and their variants**   I treat divided-bar graphs as a different type of display, not a "bar-graph variant"; unlike all other forms of bar graphs, divided bars always must display proportions. Thus they are closer to pie graphs than to the kinds of displays considered in this chapter.

chapter 6

p. 140   **highly discriminable lines**   Schutz (1961b).

p. 150   **labeling values directly**   See Culbertson & Powers (1959).

p. 154   **discriminable point symbols**   See Schutz (1961b). In addition, Chen (1982) presents evidence that symbols that differ in their topological properties (distinguished by a hole, connected components, or a closed form) are more discriminable than symbols that share topological properties (such as a filled circle and square). Cleveland & McGill (1984b) suggest that point symbols can be ordered on the basis of discriminability as follows: colors, amounts of fill, different shapes, and different letters. However, when Lewandowsky & Spence (1989) asked subjects to select two of three point clouds in a scatter plot and decide which depicted the higher correlation, they found no differences in accuracy when clouds were specified with the different symbol types. Lewandowsky & Spence did find that subjects responded more quickly with color than with shapes or fill, and required the most time when confusable letters (H, E, and F) were used as symbols. However, if letters were highly discriminable (H, Q, and X), subjects could use them faster than circles with different amounts of fill and about as well as different shapes (circle, triangle, square). Lewandowsky & Spence (1989) also found that subjects required the same amounts of time for scatter plots with 10 points and with 30 points, which suggests that subjects did not read individual symbols in their task. Keep in mind that the specific results often depend on the particular task, choices of colors, shapes, and so forth. Lewandowsky & Spence provide a table of confusions between letters, allowing the selection of highly discriminable letters to use as points (see also Geyer & DeWald, 1973).

chapter 7

p. 162   **color**   For a much more detailed discussion of color, see Travis (1991), who also offers some suggestions about how to use color to design displays that are aesthetically pleasing. A colleague told me about Travis's book as I was just finishing this one, and I wish I had seen it earlier; Tavis has done an admirable job in pulling together the relevant literature on the psychology of color and discussing it in the context of display design. I was pleased to see that he had independently formulated many of the same principles and recommendations offered in this section. This is not surprising, however, because he also based his advice on properties of the eye and brain. For additional discussion, see also Christner & Ray (1961) and Hitt (1961).

p. 162   **discriminability**   Schutz (1961b) found that his subjects could isolate points on a specific lines better if color was used to distinguish the lines. However, color helped them to find the line with the highest value at a specific point on the X axis only when the lines in a single display could be easily isolated; when the lines crossed many times, the subjects located the black-and-white ones as easily as the colored ones. In general, black-and-white displays with highly discriminable lines and symbols were almost as good as color ones.

p. 162   **"eleven colors"**   See Boynton, Fargo, Olson, & Smallman (1989) and Smallman & Boynton (1990). Travis (1991) provides precise specifications of these colors as well as detailed advice about how to produce them.

p. 162   **nine colors**   See Conover & Kraft (1954). Color can often be a useful way of making components of the content distinct (see, e.g., Aretz & Calhoun, 1982, and Casali & Gaylin, 1988), but does not always provide an advantage over black-and-white displays (see, e.g., Tullis, 1981).

p. 164   **two colors of the same brightness**   Travis (1991) points out that the oft-cited admonition not to use white letters on a yellow background (or vice versa) or black letters on a blue background (or vice versa) is in fact incorrect. The problem is that the usual way of producing those colors on a screen results in their having very similar brightnesses; if the brightnesses are varied as much as they are for other pairs of colors, it is evident that the hues are not the culprit. The lesson is to ensure that different brightnesses are used for adjacent colors.

p. 164   **blue, red, green, yellow, white**   See Travis (1991).

p. 164   **perception of brightness**   See, e.g., Cavanagh, Anstis, & Mather (1984).

p. 164   **affect the brightness**   There is a greater difference in the brightnesses of blue and yellow under fluorescent lights than under incandescent lights (blue seems brighter under fluorescent lamps); similarly, blue seems increasingly brighter than red as the lighting dims.

p. 164   **adjust the saturations**   Note that when a display composed on a computer is printed, the color of the ink usually is not exactly the same as the color on the computer screen.

p. 164   **warm colors appear "closer"**   This effect (called chromostereopsis) is a consequence of the fact that when we look straight ahead, the light passing through the lens of the eye is bent as if it were passing through a prism (with the thick side facing in, toward the nose). The effect can be exaggerated by placing such prisms in front of each eye, and can be reversed by placing the prisms so that the thick side of the wedge faces outward—in which case the light is bent so that warmer colors seem farther away. Indeed, Travis (1991) points out that the depth effect may actually reverse with low illumination (when the pupil is large, and so the eye focuses differently). For more on this fascinating phenomenon, see Allen & Rubin (1981), Held (1980, especially p. 86), Travis (1991), and Vos (1960).

p. 166 **red-green color confusion** See Pokorny, Smith, Verriest, & Pinckers (1979).

p. 166 **conventions for color** If compatibility or convention are not relevant, it often can be difficult to decide which color should stand for which entity. Travis (1991, pp. 123–124), formulates six principles of color coding that can guide you in such situations. These principles mesh nicely with those developed in this book (which are noted in parentheses following each of his principles): Colors must be discriminable (discriminability); Colors must be detectable (detectability); colors must vary in psychologically equal steps—if two colors are relatively similar, readers will believe that the things they represent are related in some way (compatibility and informative changes); colors should be meaningful—do not vary them unless they reader will know what the variations mean (informative changes and appropriate knowledge); and, colors should be aesthetically pleasing (use as few as possible and ensure that they do not vibrate or clash). Travis also suggests using no more than five different colors in a display; this recommendation also follows from the principle of limited short-term memory capacity (I do not offer it as a separate recommendation here because it is subsumed by my recommendations pertaining to the amount of content that should be displayed in a single graph). Trumbo (1981) relies on some similar ideas to speculate about the best ways to use color in maps to display more than one type of information at a time.

p. 166 **will be seen as a group** Parkin (1983; see Pinker, 1990, pp. 114–115) found that color effectively groups labels with lines, even in complex displays. Wickens & Andre (1990) provide evidence for the importance of color in helping the reader to integrate different type of information, but this evidence is less compelling for bar graphs than for exotic "object" displays (which make use of integral dimensions to pack much related information into a single display).

p. 168 **hue to represent amounts** For evidence that people have difficulty using differences in hue to extract quantitative information, see Cuff (1974) and Wainer & Francolini (1980). If practical considerations require that you use hue in this way, Travis suggests using colors that are relatively similar to construct a continuum, such as green, green-yellow, yellow, yellow-orange, orange-red, and red. This is good advice unless you want the reader to detect adjacent values at a glance; these colors are sufficiently similar that such discrimination would require effort.

p. 168 **deeper saturations/greater intensities** Based on research results, Cuff (1973) advises using these variables alone to represent quantities, holding the hue constant.

p. 170 **filled with hatching or shading** You might wonder which is better, hatching or shading. Consider the following: We want to know whether children find shape or color more important, and so we vary the shapes and colors of forms and ask the children to sort them into two piles. We could induce the children to use shape if the forms are circles vs triangles and the colors are subtly different shades of red, and could induce them to use color if the shapes are slightly different ellipses and the colors are blue and yellow. The problem is how to equate the particular values on different dimensions. This is not as easy as it might sound because context affects the way we perceive specific stimulus values; for example, the way we see a given color depends in part on the other colors that surround it (see Travis, 1991).

p. 172   **at least 30 degrees**   The estimates of the specific degree of orientation tuning of the "channels" are somewhat variable, depending on the precise technique used to measure orientation and the testing conditions (see, e.g., De Valois & De Valois, 1988, chapter 9). The estimate I offer here is somewhat conservative (some researchers suggest that individual channels respond to orientations within a range of about 20 degrees), but is consistent with relevant findings of Nothdurft (1991). Nothdurft examined the so-called pop-out phenomenon, which occurs when stimuli are sufficiently different from the background that one notices them immediately. He found that lines will seem to "pop out" from a field of other lines if their orientations differ by about 30 degrees.

p. 174   **range of spatial frequencies**   For a detailed discussion, see De Valois & De Valois (1988, chapter 6), who note that a variation of one octave (a factor of 2 to 1) to either side of a given channel will be almost totally ineffective in driving that channel.

p. 176   **depth cues**   For a review, see Kaufman (1974).

p. 178   **general impressions from three-dimensional graphs**   Casali & Gaylin (1988) found that subjects had difficulty reading specific point values from three-dimensional bars but were able to detect trends as easily as they could in standard bar or line graphs.

p. 178   **estimation of volume**   For a review, see Stevens (1974, 1975).

p. 180   **avoid see-through displays**   Huber (1987) describes an exception to this recommendation. He used moving displays on a computer screen to present multivariate scatter plots and reports that viewers can easily find outliers and clusters in such displays, as well as locate the most informative viewpoint (see also Reaven & Miller, 1979; Miller, 1985). However, he also reports (p. 451), "In general, it was our experience that only simple minded approaches will be used, and be interpretable, by anyone other than the inventor of the method."

p. 182   **use more tightly spaced grid lines**   Carter (1947) reports one of the few studies of the effectiveness of grid lines. Unfortunately, the previous recommendation ("Inner grid lines should be relatively thin and light") was not respected, and the graph with denser grid lines also lowered the contrast of the content line. Thus we cannot take at face value the finding that very frequent grid lines did not increase accuracy.

p. 184   **make some of them heavier**   See Beeby & Taylor (1973).

p. 200   **the correct visual impression**   If the overall amounts are critical, you probably should use the same scale in each panel. DeSanctis & Jarvenpaa (1985) found that graphs with different maximum values on the scales were more difficult to use in a decision-making task than graphs with identical scales; unfortunately, the graphs with different maximal values also did not use round numbers as scale values, which makes the result difficult to interpret.

chapter 8

p. 207   **varying dimensions to vary content**   Displays that do not use this technique are difficult to learn to use. For example, Chernoff faces (Chernoff, 1973) are cartoonlike faces that can have different overall shapes and different types of eyes, noses, and so on. Each value of each dimension (e.g., each type of nose) corresponds to a different value of a dependent variable. The reader must memorize by rote the arbitrary associations between the visual variations and what they represent, and later must recall these associations by rote.

p. 208   **spurious effects look significant**   To some extent, what counts as "misleading" depends on the point being made and the context. However, it is never appropriate to make a difference that is not statistically significant appear as if it were real. For example, if you measured ESP ability in five men and five women and found that women scored 1% higher on the test, this difference probably would not be statistically significant; if an apparent difference is not statistically significant, it merely reflects the sampling of random variation in the population. Thus, it would be inappropriate—whatever your point or the context—to graph the data to make this bit of random fluctuation appear as if it reflected something real about men, women, or ESP.

p. 208   **altering the axes**   I treat the X axis and the Y axis as corresponding to the independent and dependent measures respectively, but the reader should keep in mind that these roles are reversed if a horizontal bar graph is used.

p. 212   **using scales to mislead**   Cleveland, Diaconis, & McGill (1982) discuss the special case of scatter plots, finding that the variables appear more highly correlated if the range of values on the scales are increased. When the range of values is increased, the point cloud does not extend over the entire scale but instead appears as a more compact shape in the center regions of the display.

p. 214   **alter the visual impression**   For evidence that visual impression is in fact altered by aspect ratio, and a quantitative discussion of the degree to which aspect ratio and other factors affect one's impression of how steeply a content line rises, see Simcox (1984).

p. 224   **when subjects were asked**   Kosslyn & James (unpublished data, 1980).

p. 226   **salience leading to overestimation**   Kosslyn & James (unpublished data, 1980) found this to be a rather small effect, however.

p. 230   **fitting lines by eye**   Even if your intent is not to mislead, fitting a line by eye will probably produce a misleading visual impression. Mosteller, Siegel, Trapido, & Youtz (1985) asked subjects to fit lines through scatter plots by eye. They found that the average slope of these lines was closer to the slope of the "major-axis line" than to the appropriate least-squares line. The major-axis line minimizes the sum of squares of perpendicular distances, not the (correct) distances along the Y axis (see Appendix 1 for a discussion of least-squares fits).

p. 232   **visualization-of-animals study**   For details, see Kosslyn (1976); for an overview of this sort of research, see Kosslyn (1983).

chapter 9

p. 238 **charts, diagrams, and maps** Winn (1987) characterizes charts, graphs, and diagrams in a similar way to that offered here, but his "charts" can be entirely textual. According to the present characterization, a chart can include only text if spatial relations are used to convey additional information about the material. For example, an organizational chart might have one name at the top, two in a row directly under that, six under each of those names, and so on. If the reader understands that this is an organizational chart, it can be read as specifying a hierarchy. Moxley (1983) distinguished among three types of diagrams, some of which are charts by the present characterization; consistent with the present approach, Fleming & Levie (1978) emphasize the role of visual representations of relations (such as inclusion, subordination and so on) in charts. Bertin (1973) offers a very elaborate discussion of such matters (see Wainer & Francolini, 1980, for a discussion of the rudiments of Bertin's ideas, and Kosslyn, 1985, for a review of Bertin's book). Tufte (1983, 1990) provides many instructive illustrations.

p. 238 **symbolic content** See, e.g., Winn (1987).

p. 240 **compatible layout** For related ideas, see Fleming & Levie (1978) and Winn (1987).

p. 240 **flow chart . . . should start at the left and work to the right** The convention in computer science is to start a flow chart at the top and progress downward. This convention was adopted because the lines of a computer program are listed down a page, and each successive component of a flow chart corresponds to a successive unit of the program's code.

p. 246 **use a map** For more information about when to use and how to design maps, see Fisher (1982) and Monmonier (1991).

p. 248 **draw a bar or similar symbol** Maps and visual tables can be combined, creating a hybrid format. Cleveland & McGill (1984a), Cleveland (1985), and Dunn (1987) have argued that a "framed rectangle" chart is more effective than simply making regions of a map darker to indicate greater amounts at that location. A framed rectangle chart has a constant-sized bar (the frame) at or near the center of each location, and a bar is presented within the frame. The frame has horizontal ticks, allowing the reader to estimate the height of the bar. Dunn (1988) tested the effectiveness of such displays. He first prepared maps that depicted murder rates in the United States; in one version, a framed rectangle was imposed on each state, and in the other different amounts of fill were placed in each state. The subjects could in fact read quantities from the framed rectangle display more accurately than from the filled map (such displays are termed choropleth maps). We can extend the basic concept of a framed rectangle display, replacing bars with pictures. For example, in a map indicating population in different regions, the bars could be replaced by drawings of people sitting on each others' shoulders, with each person representing a fixed unit of population. In such cases the map serves to label the content elements of the visual table.

bibliography

ALDRICH, F. K., & PARKIN, A. J. (1987). Tangible line graphs: An experimental investigation of three formats using capsule paper. *Human Factors, 29,* 301–309.

ALLEN, R. C., & RUBIN, M. L. (1981). Chromostereopsis. *Survey of Ophthalmology, 26,* 22–27.

ANDERSON, J.R. (Ed.) (1981). *Cognitive skills and their acquisition.* Hillsdale, NJ: Erlbaum.

ANDERSON, J. R. (1990). *Cognitive psychology and its implications.* New York: W. H. Freeman.

ARETZ, A. J., & CALHOUN, G. L. (1982). Computer generated pictorial stores management displays for fighter aircraft. *Proceedings of the Human Factors Society 26th Annual Meeting* (pp. 455–459). Santa Monica, CA: Human Factors Society.

BADDELEY, A. D. (1986). *Working memory.* Oxford: Oxford University Press.

BAIRD, J. C. (1970). *Psychophysical analysis of visual space.* New York: Pergamon Press.

BARNETT, B. J., & WICKENS, C. D. (1988). Display proximity in multicue information integration: The benefits of boxes. *Human Factors, 30,* 15–24.

BEEBY, A. W., & TAYLOR, H. P. J. (1973). How well can we use graphs? *Communication of Scientific and Technical Information, 17,* 7–11.

BERINGER, D. B. (1987). Peripheral integrated status displays. *Displays, 1,* 33–36.

BERTIN, J. (1983). (W. J. Berg, Trans). *Semiology of graphs.* Madison, WI: University of Wisconsin Press.

BICKEL, P. J., HAMMEL, E. A., & O'CONNELL, J. W. (1975). Sex bias in graduate admissions: Data from Berkeley. *Science, 187,* 398-404.

BLOOMER, C. M. (1990). *Principles of visual perception (second ed.).* New York: Design Press.

BOYNTON, R. M., FARGO, L., OLSON, C. X., & SMALLMAN, H. S. (1989). Category effects in color memory. *Color Research and Application, 14,* 229–234.

BRINTON, W. C. (1916). *Graphic methods of presenting facts.* New York: Engineering Magazine.

BROWN, R. L. (1985). Methods for graphic representation of simulated data. *Ergonomics, 28,* 1439–1454.

BRYANT, P. E., & SOMERVILLE, S. C. (1986). The spatial demands of graphs. *British Journal of Psychology, 77,* 187–197.

CARSWELL, C. M. (1992). Choosing specifiers: An evaluation of the basic tasks model of graphical perception. *Human Factors, 34,* 535–554.

CARSWELL, C. M., & WICKENS, C. D. (1987a). Information integration and the object display: An interaction of task demands and display superiority. *Ergonomics, 30,* 511–528.

CARSWELL, C. M., & WICKENS, C. D. (1987b). Objections to objects: Limitations of human performance in the use of iconic graphics. In L. S. Mark, J. S. Warm, & R. L. Huston (Ed.), *Ergonomics and human factors: Recent research* (pp. 253–260). New York: Springer-Verlag.

CARSWELL, C. M., & WICKENS, C. D. (1988). *Comparative graphics: History and applications of perceptual integrality theory and the proximity compatibility hypothesis.* Technical Report. Aviation Research Lab, Institute of Aviation, University of Illinois.

CARSWELL, C. M., & WICKENS, C. D. (1990). The perceptual interaction of graphic attributes: Configurality, stimulus homogeneity, and object integration. *Perception and Psychophysics, 47,* 157–168.

CARTER, L. F. (1947). An experiment on the design of tables and graphs used for presenting numerical data. *Journal of Applied Psychology, 31,* 640–650.

CASALI, J. G., & GAYLIN, K. B. (1988). Selected graph design variables in four interpretation tasks: A microcomputer-based pilot study. *Behaviour and Information Technology, 7,* 31–49.

CASEY, E. J., & WICKENS, C. D. (1986). *Visual display representation of multidimensional systems.* Champaign, IL: Cognitive Psychophysiology Laboratory, University of Illinois.

CASNER, S., & LARKIN, J. H. (1989). Cognitive efficiency considerations for good graphic design. *Proceedings of the Cognitive Science Society.* Hillsdale, NJ: Erlbaum Associates.

CAVANAGH, P., ANSTIS, S., & MATHER, G. (1984). Screening for color blindness using optokinetic nystagmus. *Investigative Ophthalmology and Visual Science, 25,* 463–466.

CHAMBERS, J. M., CLEVELAND, W. S., KLEINER, B., & TUKEY, P. A. (1983). *Graphical methods for data analysis.* Belmont, CA: Wadsworth.

CHEN, L. (1982). Topological structure in visual perception. *Science, 218,* 699–700.

CHERNOFF, H. (1973). The use of faces to represent points in k-dimensional space graphically. *Journal of the American Statistical Association, 68,* 361–368.

CHERNOFF, H., & RIZVI, H. M. (1975). Effect on classification error of random permutations of features in representing multivariate data by faces. *Journal of the American Statistical Association, 70,* 548–554.

CHRISTNER, C. A., & RAY, H. W. (1961). An evaluation of the effect of selected combinations of target and background coding on map-reading performance— Experiment V. *Human Factors, 3,* 131–146.

CLARK, H. H., & CLARK, E. V. (1977). *Psychology and language: An introduction to psycholinguistics.* New York: Harcourt, Brace, Jovanovich.

CLEMENT, D. E. (1978). Perceptual structure and selection. In E. C. Carterette & M. P. Friedman (Ed.), *Handbook of perception, vol. 9* (pp. 49–84). New York: Academic Press.

CLEVELAND, W. S. (1985). *The elements of graphing data.* Monterey, CA: Wadsworth.

CLEVELAND, W. S. (1987). Research in statistical graphics. *Journal of the American Statistical Association, 82,* 419–423.

CLEVELAND, W. S., DIACONIS, P., & McGILL, R. (1982). Variables on scatterplots look more highly correlated when the scales are increased. *Science, 216,* 1138–1141.

CLEVELAND, W. S., HARRIS, C. S., & McGILL, R. (1982). Judgments of circle sizes on statistical maps. *Journal of the American Statistical Association, 77,* 541–547.

CLEVELAND, W. S., & McGILL, R. (1984a). Graphical perception: Theory, experimentation, and application to the development of graphical methods. *Journal of the American Statistical Association, 79,* 531–554.

CLEVELAND, W. S., & McGILL, R. (1984b). The many faces of a scatterplot. *Journal of the American Statistical Association, 79,* 807–822.

CLEVELAND, W. S., & McGILL, R. (1985). Graphical perception and graphical methods for analyzing scientific data. *Science, 229,* 828–833.

CLEVELAND, W. S., & McGILL, R. (1986). An experiment in graphical perception. *International Journal of Man-Machine Studies, 25,* 491–500.

CLEVELAND, W. S., & McGILL, R. (1987). Graphical perception: The visual decoding of quantitative information on graphical displays of data. *Journal of the Royal Statistical Society, 150 (Series A, Part 3),* 192–229.

COFFEY, J. L. (1961). A comparison of vertical and horizontal arrangements of alpha-numeric material—Experiment 1. *Human Factors, 3,* 93–98.

CONDOVER, D. W., & KRAFT, C. L. (1954). *The use of color in coding displays.* Wright-Patterson Air Force Base, Ohio: Wright Air Development Center.

COOPER, W. E., & ROSS, J. R. (1975). *World order. Notes from the parasession on functionalism.* Chicago, IL: Chicago Linguistic Society.

COURY, B. G., BOULETTE, M. D., & SMITH, R. A. (1989). Effect of uncertainty and diagnosticity on classification of multidimensional data with integral and separable displays of system status. *Human Factors, 31,* 551–569.

CROXTON, F. E., & STEIN, H. (1932). Graphic comparisons by bars, squares, circles, and cubes. *Journal of the American Statistical Association, 27,* 54–60.

CROXTON, F. E., & STRYKER, R. E. (1927). Bar charts versus circle diagrams. *Journal of the American Statistical Association, 22,* 473–482.

CUFF, D. J. (1973). Colour on temperature maps. *Cartographic Journal, 10,* 17–21.

CULBERTSON, H. M., & POWERS, R. D. (1959). A study of graph comprehension difficulties. *Audio Visual Communication Review, 7,* 97–100.

CURCIO, F. R. (1987). Comprehension of mathematical relationships expressed in graphs. *Journal for Research in Mathematics Education, 18,* 382–393.

DE VALOIS, R. L., & DE VALOIS, K. K. (1988). *Spatial vision.* New York: Oxford University Press.

DESANCTIS, G. (1984). Computer graphics as decision aids: Direction for research. *Decision Science, 15,* 463–487.

DESANCTIS, G., & JARVENPAA, S. L. (1985). An investigation of the "tables versus graphs" controversy in a learning environment. In L. Gallegos, R. Welke, & J. Wetherbe (Ed.), *Proceedings of the 6th international conference on information systems* (pp. 134–144). Indianapolis, IN: Research and Education Information Systems.

DeSoete, G., & DeCorte, W. (1985). On the perceptual salience of features of Chernoff faces for representing multivariate data. *Applied Psychological Measurement, 9*, 275–280.

Dodwell, P. C. (1975). Perceptual structure and selection. In E. C. Carterette, & M. P. Friedman (Ed.), *Handbook of perception, vol. 5* (pp. 267-300). New York: Academic Press.

Dunn, R. (1987). Variable-width framed rectangle charts for statistical mapping. *American Statistician, 41*, 153–156.

Dunn, R. (1988). Framed rectangle charts or statistical maps with shading. *American Statistician, 42*, 123-129.

Eells, W. C. (1926). The relative merits of circles and bars for representing component parts. *Journal of the American Statistical Association, 21*, 119–132.

Ehrenberg, A. S. C. (1975). *Data reduction: Analyzing and interpreting statistical data*. New York: Wiley.

Ericsson, K. A., Chase, W. G., & Faloon, S. (1980). Acquisition of a memory skill. *Science, 208*, 1181–1182.

Feinberg, S. E. (1979). Graphical methods in statistics. *American Statistician, 33*, 165–178.

Fisher, H. T. (1982). *Mapping information*. Cambridge, MA: Abt Books.

Fleming, M. L., & Levie, W. H. (1978). *Instructional message design: Principles from the behavioral sciences*. Englewood Cliffs, NJ: Educational Technology Publications.

Frenckner, K. (1990). *Legibility of continuous text on computer screens—a guide to the literature (TRITA-NA P9010, IPLab-25)*. Stockholm, Sweden: Royal Institute of Technology

Frisby, J. P. (1980). *Seeing: Illusion, brain, and mind*. New York: Oxford University Press.

Games, P.A., & Klare, G. R. (1967). *Elementary statistics: Data analysis for the behavioral sciences*. New York: McGraw-Hill.

Garner, W. R. (1970). The stimulus in information processing. *American Psychologist, 25*, 350–358.

Garner, W. R. (1974). *The processing of information and structure*. Hillsdale, NJ: Erlbaum.

Garner, W. R. (1976). Interaction of stimulus dimensions in concept and choice processes. *Cognitive Psychology, 8*, 98–123.

Geyer, L. H., & DeWald, C. G. (1973). Feature lists and confusion matrices. *Perception and Psychophysics, 14*, 471–482.

Glass, A. R., & Holyoak, K. J. (1986). *Cognition (second ed.)*. New York: Random House.

Goldsmith, T. E., & Schvaneveldt, R. W. (1985). Facilitating multiple-cue judgments with integral information displays. In J. Thomas, & M. Schneider (Ed.), *Human factors in computer systems*. Norwood, NJ: Ablex.

Gould, J. D., Alfaro, L., Finn, R., Haupt, B., & Minuto, A. (1987). Reading from CRT displays can be as fast as reading from paper. *Human Factors, 29*, 497–517.

GRAHAM, J. L. (1937). Illusory trends in the observation of bar graphs. *Journal of Experimental Psychology, 20,* 597–608.

GREGORY, R. L. (1966). *Eye and brain: The psychology of seeing.* New York: McGraw-Hill.

GREGORY, R. L. (1970). *The intelligent eye.* London: Weidenfeld and Nicholson.

GRICE, H. P. (1975). Logic and conversation. In P. Cole, & J. L. Morgan (Ed.), *Syntax and semantics, vol. 3: Speech acts* (pp. 41–58). New York: Seminar Press.

HELD, R. (1980). The rediscovery of adaptability in the visual system: Effects of extrinsic and intrinsic chromatic dispersion. In C. S. Harris (Ed.), *Visual coding and adaptability* (pp. 69–94). Hillsdale, NJ: Erlbaum.

HITT, W. D. (1961). An evaluation of five different abstract coding methods— Experiment IV. *Human Factors, 3,* 120–130.

HOCHBERG, J. E. (1964). *Perception.* Englewood Cliffs, NJ: Prentice-Hall.

HOLMES, N. (1984). *Designer's guide to creating charts and diagrams.* New York: Watson-Guptill.

HOTOPF, W. H. N. (1966). The size-constancy theory of visual illusions. *British Journal of Psychology, 57,* 307–318.

HOTOPF, W. H. N., & HIBBERD, M. C. (1989). The role of angles in inducing misalignment in the Poggendorf figure. *Quarterly Journal of Experimental Psychology, 41A,* 355–383.

HOWARD, I. P. (1982). *Human visual orientation.* New York: John Wiley.

HUBER, P. J. (1987). Experiences with three-dimensional scatterplots. *Journal of the American Statistical Association, 82,* 448–453.

HUFF, D., and GEIS, I. (1954). *How to lie with statistics.* New York: W. W. Norton.

JACOB, R. J. K., EGETH, H. E., & BEVAN, W. (1976). The face as a data display. *Human Factors, 18,* 189–200.

JARVENPAA, S. L., & DICKSON, G. W. (1988). Graphics and managerial decision making: Research based guidelines. *Communications of the ACM, 31,* 764–774.

JENKS, C. F., & KNOS, D. S. (1961). The use of shading patterns in graded series. *Annals of the Association of American Geographers, 51,* 316–334.

JOHNSON, J. R., RICE, R. R., & ROEMMICH, R. A. (1980). Pictures that lie: The abuse of graphs in annual reports. *Management Accounting, 62,* 50–56.

JOLICOEUR, P. (1990). Identification of disoriented objects: A dual-systems theory. *Mind and Language, 5,* 387–410.

JONES, P., & WICKENS, C. D. (1986). *The display of multivariate information: The effects of auto- and cross-correlation, display format, and reliability.* Champaign, IL: Cognitive Psychophysiology Laboratory, University of Illinois.

JULEZ, B. (1980). Spatial-frequency channels in one-, two-, and three-dimensional vision: Variations on a theme by Bekesy. In C. S. Harris (Ed.), *Visual coding and adaptability* (pp. 263–316). Hillsdale, NJ: Erlbaum.

KAHNEMAN, D. (1973). *Attention and effort.* Englewood Cliffs, NJ: Prentice-Hall.

KAUFMAN, L. (1974). *Sight and mind: An introduction to visual perception.* New York: Oxford.

KOLATA, G. (1984). The proper display of data. *Science, 226*, 156–157.

KOSSLYN, S. M. (1976). Using imagery to retrieve semantic information: A developmental study. *Child Development, 47*, 434–444.

KOSSLYN, S. M. (1983). *Ghosts in the mind's machine.* New York: W. W. Norton.

KOSSLYN, S. M. (1985). Graphics and human information processing: A review of five books. *Journal of the American Statistical Association, 80*, 499–512.

KOSSLYN, S. M. (1989). Understanding charts and graphs. *Applied Cognitive Psychology, 3*, 185–225.

KOSSLYN, S. M., AND KOENIG, O. (1992). *Wet mind: The new cognitive neuroscience.* New York: Free Press.

LEDERMAN, S. J., & CAMPBELL, J. I. (1982). Tangible graphs for the blind. *Human Factors, 24*, 85–100.

LEDERMAN, S. J., & CAMPBELL, J. I. (1983). Tangible line graphs: An evaluation and some systematic strategies for exploration. *Journal of Visual Impairment and Blindness, 77*, 108–112.

LEGGE, G. E., GU, Y., & LUEBKER, A. (1989). Efficiency of graphical perception. *Perception and Psychophysics, 46*, 365–374.

LEWANDOWSKY, S., & SPENCE, I. (1989). Discriminating strata in scatterplots. *Journal of the American Statistical Association, 84*, 682–688.

LUCAS, H. C. J. (1981). An experimental investigation of the use of computer-based graphics in decision making. *Management Science, 27*, 757–768.

MACDONALD-ROSS, M. (1977). How numbers are shown: A review of research on the presentation of quantitative data in texts. *Audio-Visual Communication Review, 25*, 359–409.

MACGREGOR, D., & SLOVIC, P. (1986). Graphic representation of judgmental information. *Human-Computer Interaction, 2*, 179–200.

MACLEAN, I. E., & STACEY, B. G. (1971). Judgment of angle size: An experimental appraisal. *Perception and Psychophysics, 9*, 499–504.

MACLEOD, V. M. (1991). Half a century of research on the Stroop effect: An integrative review. *Psychological Bulletin, 109*, 163–203.

MEIHOEFER, H. J. (1969). The utility of the circle as an effective cartographic symbol. *Canadian Cartographer, 6*, 105–117.

MEIHOEFER, H. J. (1973). The visual perception of the circle in thematic maps: Experimental results. *Canadian Cartographer, 10*, 63–84.

MILLER, G. A. (1956). The magical number seven, plus or minus two: Some limits on our capacity for processing information. *Psychological Review, 63*, 81–97.

MILLER, R. G. (1985). Discussion of "Projection pursuit," by P. J. Huber. *Annals of Statistics, 13*, 510–513.

MILROY, R., & POULTON, E. C. (1978). Labeling graphs for improved reading speed. *Ergonomics, 21*, 55–61.

MISHKIN, M., & APPENZELLER, T. (1987). The anatomy of memory. *Scientific American, 256*, 80–89.

MOKROS, J. R., & TINKER, R. F. (1987). The impact of microcomputer-based labs on children's ability to interpret graphs. Special issue: *Cognitive consequences of technology in science education. Journal of Research in Science Teaching, 24,* 369–383.

MONMONIER, M. (1991). *How to lie with maps.* Chicago, IL: University of Chicago Press.

MORIARITY, S. (1979). Communicating financial information through multi-dimensional graphics. *Journal of Accounting Research, 17,* 205–224.

MOSTELLER, F., SIEGEL, A. F., TRAPIDO, E., & YOUTZ, C. (1985). Fitting straight lines by eye. In D. C. Hoaglin, F. Mosteller, & J. W. Tukey (Ed.), *Exploring data tables, trends, and shapes* (pp. 225–240). New York: John Wiley.

MOXLEY, R. (1983). Educational diagrams. *Instructional Science, 12,* 147–160.

NAWROCKI, L. H. (1973). *Graphic versus tote display of information in a simulated tactical operations system (Tech. Res. Note No. 243).* Washington, DC: Army Research Institute for the Behavioral and Social Sciences.

NEISSER, U. (1967). *Cognitive psychology.* New York: Appleton-Century-Crofts.

NEISSER, U. (1976). *Cognition and reality.* San Francisco: W. H. Freeman.

NEURATH, A. (1974). Isotype. *Instructional Science, 3,* 127-150.

NEWELL, A. (1990). *Unified theories of cognition.* Cambridge, MA: Harvard University Press.

NOTHDURFT, H. C. (1991). Texture segmentation and pop-out from orientation contrast. *Vision Research, 31,* 1073–1078.

OSHERSON, D., KOSSLYN, S. M., & HOLLERBACH, J. (Eds.) (1990). *An invitation to cognitive science: Visual cognition and action, vol. 2.* Cambridge, MA: MIT Press.

OSHERSON, D. N., & SMITH, E. E. (Eds.) (1990). *Thinking: An invitation to cognitive science, vol. 3.* Cambridge, MA: MIT Press.

PATTERSON, D. G., & TINKER, M. A. (1940). *How to make type readable: A manual for typographers, printers and advertisers.* New York: Harper & Brothers.

PETERSON, L. V., & SCHRAMM, W. (1954). How accurately are different kinds of graphs read? *Audio-Visual Communication Review, 2,* 178–189.

PETERSON, R. J., BANKS, W. W., & GERTMAN, D. I. (1982). Performance-based evaluation of graphic displays for nuclear power plant control rooms. *Proceedings of the Conference on Human Factors in Computer Systems* (pp. 182–209). New York: ACM.

PHILLIPS, R. J., COE, B., KONO, E., KNAPP, J., BARRETT, S., WISEMAN, G., & EVELEIGH, P. (1990). An experimental approach to the design of cartographic symbols. *Applied Cognitive Psychology, 4,* 485–497.

PINKER, S. (1990). A theory of graph comprehension. In R. Freedle (Ed.), *Artificial intelligence and the future of testing* (pp. 73–126). Hillsdale, NJ: Lawrence Erlbaum Associates.

PINKER, S., & BIRDSONG, D. (1979). Speakers' sensitivity to rules of frozen word order. *Journal of Verbal Learning and Verbal Behavior, 18,* 497–508.

PLAYFAIR, W. (1786). *The commercial and political atlas.* London: Corry.

POKORNY, J., SMITH, V. C., VERRIEST, G., & PINCKERS, A. J. L. G. (1979). *Congenital and acquired colour vision defects.* New York: Grune and Stratton.

297

POMERANTZ, J. R. (1981). Perceptual organization in information processing. In M. Kubovy, & J. R. Pomerantz (Ed.), *Perceptual organization* (pp. 141–180). Hillsdale, NJ: Erlbaum.

POSNER, M. I. (1978). *Chronometric explorations of mind.* Hillsdale: Lawrence Erlbaum.

POULTON, E. C. (1985). Geometric illusions in reading graphs. *Perception and Psychophysics, 37,* 543–548.

REAVEN, G. M., & MILLER, R. G. (1979). An attempt to define the nature of chemical diabetes using a multidimensional analysis. *Diabetologia, 16,* 17–24.

REED, S. K., & JOHNSEN, J. A. (1975). Detection of parts in patterns and images. *Memory and Cognition, 3,* 569–575.

ROBINSON, J. O. (1972). *The psychology of visual illusion.* London: Hutchinson.

SANDERSON, P. M., FLACH, J. M., BUTTIGIEG, M. A., & CASEY, E. J. (1989). Object displays do not always support better integrated task performance. *Human Factors, 31,* 183–198.

SCHIANO, D. J., & TVERSKY, B. (1992). Structure and strategy in encoding simplified graphs. *Memory & Cognition, 20,* 12–20.

SCHMID, C. F. (1954). *Handbook of graphic presentation.* New York: Ronald.

SCHUTZ, H. G. (1961a). An evaluation of formats for graphic trend displays—Experiment II. *Human Factors, 3,* 99–107.

SCHUTZ, H. G. (1961b). An evaluation of methods for presentation of graphic multiple trends—Experiment III. *Human Factors, 31,* 108–119.

SCHWARTZ, D. R. (1984). *Optional stopping performance under graphic and numeric CRT formatting (Tech. Rep. No. 84-1).* Houston, TX: Department of Psychology, Rice University.

SHANNON, C. E. (1948). A mathematical theory of communication. *Bell System Technical Journal, 27,* 379–423, 623–656.

SIMCOX, W. A. (1983). *A perceptual analysis of graphic information processing.* Unpublished Ph.D. dissertation, Tufts University.

SIMCOX, W. A. (1984). A method for pragmatic communication in graphic displays. *Human Factors, 26,* 483–487.

SIMKIN, D., & HASTIE, R. (1987). An information processing analysis of graph perception. *Journal of the American Statistical Association, 82,* 454–465.

SMALLMAN, H. S., & BOYTON, R. M. (1990). Segregation of basic colors in an information display. *Journal of the Optical Society of America, A7,* 1985–1994.

SMITH, S. L. (1979). Letter size and legibility. *Human Factors, 21(b),* 661–670.

SPENCE, I. (1990). Visual psychophysics of simple graphical elements. *Journal of Experimental Psychology: Human Perception and Performance, 16,* 683–692.

SPENCE, I., & LEWANDOWSKY, S. (1991). Displaying proportions and percentages. *Applied Cognitive Psychology, 5,* 61–77.

SPOEHR, K. T., & LEHMKUHLE, S. W. (1982). *Visual information processing.* San Francisco: W. H. Freeman.

STEVENS, S. S. (1974). Perceptual magnitude and its measurement. In E.C. Carterette & M.P. Friedman (Eds.), *Handbook of perception, vol. 2* (pp. 361–389). New York: Academic Press.

STEVENS, S. S. (1975). *Psychophysics*. New York: John Wiley.

STOCK, D., & WATSON, C. J. (1984). Human judgment accuracy, multidimensional graphics, and human versus models. *Journal of Accounting Research, 22,* 192–206.

TEGHTSOONIAN, M. (1965). The judgment of size. *American Journal of Psychology, 78,* 392–402.

TINKER, M. A. (1963). *Legibility of print*. Ames, IA: Iowa State University Press.

TRAVIS, D. (1991). *Effective color displays: Theory and practice*. New York: Academic Press.

TRUMBO, B. E. (1981). A theory for coloring bivariate statistical maps. *American Statistician, 35,* 220–226.

TUFTE, E. R. (1983). *The visual display of quantitative information*. Cheshire, CT: Graphics Press.

TUFTE, E. R. (1990). *Envisioning information*. Cheshire, CT: Graphics Press.

TUKEY, J. W. (1972). Some graphic and semigraphic displays. In T. A. Bancroft (Ed.), *Statistical papers in honor of George W. Snedecor* (pp. 293–316). Ames, IA: Iowa State University Press.

TUKEY, J. W. (1977). *Exploratory data analysis*. Reading, MA: Addison-Wesley.

TULLIS, T. S. (1981). An evaluation of alphanumeric, graphic, and colour information displays. *Human Factors, 23,* 541–550.

TVERSKY, B., & SCHIANO, D. J. (1989). Perceptual and cognitive factors in distortions in memory for graphs and maps. *Journal of Experimental Psychology: General, 118,* 387–398.

UMANATH, N. S., & SCAMELL, R. W. (1988). An experimental evaluation of the impact of data display format on recall performance. *Communications of the ACM, 31,* 562–570.

VERNON, M. D. (1946). Learning from graphic material. *British Journal of Psychology, 36,* 145–158.

VON HUHN, R. (1927). Further studies in the graphic use of circles and bars. *Journal of the American Statistical Association, 22,* 31–36.

VOS, J. J. (1960). Some new aspects of color stereoscopy. *Journal of the Optical Society of America, 50,* 785–790.

WAINER, H. (1979). *The Wabbit: An alternative icon for multivariate data display (BSSR Technical report, 547–792)*. Washington, DC: Bureau of Social Science Research

WAINER, H., & FRANCOLINI, C. M. (1980). An empirical inquiry concerning human understanding of two-variable color maps. *American Statistician, 34,* 81–93.

WAINER, H., & THISSEN, D. (1981). Graphical data analysis. *Annual Review of Psychology, 32,* 191–241.

WAKIMOTO, K. (1977). A trial of modification of the face graph proposed by Chernoff: Body graph. *Quantitative Behavioral Science, Kodo Keiryogaku, 4,* 67–73.

WASHBURNE, J. N. (1927). An experimental study of various graphic, tabular, and textual methods of presenting quantitative material. *Journal of Educational Psychology, 18,* 361–376.

WICKENS, C. D., & ANDRE, A. D. (1990). Proximity compatibility and information display: Effects of color, space, and objectness on information integration. *Human Factors, 32,* 61–77.

WINN, W. D. (1983). Perceptual strategies used with flow diagrams having normal and unanticipated formats. *Perceptual and Motor Skills, 57,* 751–762.

WINN, B. (1987). Charts, graphs, and diagrams in educational materials. In D. M. Willows, & H. A. Houghton (Ed.), *The psychology of illustration, vol. 1: Basic research.* New York: Springer-Verlag.

ZACHRISSON, B. (1965). *Studies in the legibility of printed text.* Stockholm, Sweden: Almqvist & Wiksell.

sources of data and figures

All the figures that appear in this book are original drawings, prepared by Network Graphics, Vantage Art, and Megan Higgins. Many use fictional data; others, listed below, were adapted from the following sources:

4   "The recognition of faces" by L. D. Harmon, 1973, *Scientific American, 229*, p. 75.

6   *left*, Consensus Economics, Inc., London, cited in *The Economist*, 15 December 1990, p. 99; *right, The Economist*, 4 May 1991, p. 103.

15   "Social policy and recent fertility changes in Sweden" by J. M. Hoem, *Population and Development Review*, December 1990, cited in *The Economist*, 13 April 1991, p. 47.

20   Mishkin, M., Ungerleider, L. G., & Macko, K. A. (1983). Object vision and spatial vision: Two cortical pathways. *Trends in Neurosciences, 6*, 414–417.

25   U.S. Federal Reserve and Bureau of Labor Statistics, 2.5 cited in *The Economist*, 1 December 1990, p. 94.

27   *top*, DRI/McGraw-Hill, cited in *The Economist*, 8 June 1991, Survey p. 23; *bottom*, ASEA Annual Report, 1989.

29   U.S. Forest Service, cited in *The Economist*, 22 June 1991, p. 23.

37   *bottom*, Arthur Andersen, Inc., cited in *The Economist*, 17 August 1991, p. 67.

41   *top*, Inland Revenue (U.K.), cited in *The Economist*, 14 March 1992, p. 73.

43   *top*, Kosslyn, S. M., Sokolov, M. A. & Chen, J. C. (1989). THe lateralization of BRIAN: A computational theory and model of visual hemispheric specialization. In D. Klahr and K. Kotovsky (Eds.), *Complex information processing: The impact of Herbert H. Simon*. Hillsdale, NJ: Erlbaum.

45   DRI/McGraw-Hill, cited in *The Economist*, 8 June 1991, p. 23.

47   *bottom, Consumer Reports*, April 1979, and U.S. Environmental Protection Agency, cited in Chambers, Cleveland, Kleiner, & Tukey (1983, p. 139).

49   *Consumer Reports*, April 1979, and U.S. Environmental Protection Agency, cited in Chambers et al. (1983, p. 133).

53   *bottom, The Scientist*, 5 March 1990, p. 10.

57   *bottom, Petrology* (p. 295) by E. G. Ehlers and H. Blatt, 1982, New York: W. H. Freeman.

59   *bottom*, Cleveland (1985, p. 276).

71   *The New York Times*, 20 October 1991, sec. 8, p. 2.

77   *bottom*, The Royal Society, cited in *The Economist*, 18 May 1991, p. 63.

79   Leo J. Shapiro & Associates, cited in *The Wall Street Journal*, 4 October 1991, p. B1.

81   *The computational brain: Models and methods on the frontiers of computational neuroscience* (p. 428) by P. S. Churchland and T. J. Sejnowski, 1992, Cambridge, MA: MIT Press.

83   *top*, U.S., Immigration and Naturalization Service, 3.16 cited in *The Economist*, 11 May 1991, p. 18; *bottom*, Arthur Andersen, Inc., cited in *The Economist*, 17 August 1991, p. 67.

85   *top*, The Conference Board, cited in *The Economist*, 18 May 1991, p. 74.

89   *top*, *The Wall Street Journal*, 21 May 1990, p. A4; 3.23 *bottom*, Datastream, U.S. Department of Commerce, and Yamaichi Research Institute, cited in *The Economist*, 31 August 1991, p. 84.

91   Metropolitan Life Insurance Company height and weight tables, cited in *The psychology of eating and drinking: An introduction* (p. 172) by A. W. Logue, 1991, New York: W. H. Freeman.

93   *top*, *The Economist*, 31 August 1991, p. 17.

97   *bottom*, Department of the Environment (U.K.), cited in *The Economist*, 20 April 1991, p. 56.

99   *top*, *The Economist*, 14 March 1992, p. 67.

103   World Resources Institute, cited in *Ecology and the politics of scarcity revisited* (p. 74), W. Ophuls and A. S. Boyan, Jr., 1992, New York: W. H. Freeman.

113   *top*, *Cultural Trends, 1990*, cited in *The Economist*, 15 December 1990, p. 55.

115   *middle*, National Association of Manufacturers and Bank of England, cited in *The Economist*, 12 January 1991, p. 60.

121   *top*, *Jesse Meyers' Beverage Digest*, cited in *Newsweek*, 19 March 1990, p. 38.

123   *top*, *The Economist*, 23 November 1991, Survey p. 14; 5.4 *bottom*, *The why and how of home horticulture*, second ed. (p. 480), by D. R. Bienz, 1993, New York: W. H. Freeman.

125   *top*, *Jesse Meyers' Beverage Digest*, cited in *Newsweek*, 19 March 1990, p. 38.

127   Greater Tampa Association of Realtors, St. Petersburg Suncoast Association of Realtors, and Greater Clearwater Board of Realtors, cited in *Tampa Tribune*, 29 May 1991.

129   *Cultural Trends, 1990*, cited in *The Economist*, 15 December 1990, p. 55.

131   *bottom*, *The evolution of Japanese direct investment in Europe* by S. Thomsen and P. Nicolaides, 1991, New York: Harvester Wheatsheaf, and U.S. Department of Commerce, cited in *The Economist*, 20 April 1991, p. 65.

135   *top*, *The beginnings of social understanding* (p. 48) by J. Dunn, 1988, Cambridge, MA: Harvard University Press, cited in *The development of children*, second ed (p. 377), by M. Cole and S. R. Cole, 1993, New York: W. H. Freeman.

137   *The sports encyclopedia: pro football* by David Neft, Richard Cohen and Rick Korch, 1992, New York: St. Martin's Press.

141   *top*, Nielsen Media Research, cited in *Newsweek*, 26 March 1990, p. 58; *bottom*, Schutz (1961b).

147   Dun & Bradstreet and Department of Employment (U.K.), cited in *The Economist*, 5 January 1991, p. 42.

149   *top*, *The Economist*, 23 November 1991, p. 59; *bottom*, Greater Tampa Association of Realtors, St. Petersburg Suncoast Association of Realtors, and Greater Clearwater Board of Realtors, cited in *Tampa Tribune*, 29 May 1991.

153 Department of the Environment (U.K.), cited in *The Economist*, 20 April 1991, p. 56.

165 *top, The new archaeology and the ancient Maya* (p. 130) by J. A. Sabloff, 1990, New York; Scientific American Library; *bottom*, "Development of moral judgement: A longitudinal study of males and females by C. Holstein, 1976, *Child Development*, 47, pp. 51–61, cited in Cole & Cole (1993, p. 629).

173 *bottom*, King's Fund Institute, cited in *The Economist*, 22 February 1992, p. 49.

179 *bottom*, Baseball Commissioner, cited in *The Economist*, 13 April 1991, p. 67.

181 *middle, Fortune* and *Forbes* magazines, cited in *Designer's guide to creating charts and diagrams* (p. 46) by N. Holmes, 1984, New York: Watson-Guptill.

183 *top*, Comité Interprofessionnal du Vin de Champagne, cited in *The Economist*, 1 December 1990, p. 69.

185 *top*, U.S. Department of Commerce and Bureau of the Census, cited in *The Economist*, 6 July 1991, p. 26; *bottom, The economy of nature*, third ed. (p. 71), by R. E. Ricklefs, 1993, New York: W. H. Freeman.

187 *bottom, The New York Times*, 17 February 1992, p. A12.

189 SIPRI, cited in *The Economist*, 6 July 1991, p. 33.

193 *top, Infancy: Its place in human development* by J. Kagan, R. B. Kearsley, and P. Zelazo, 1978, Cambridge, MA: Harvard University Press, cited in Cole & Cole (1993, p. 236); *bottom, The New York Times*, 26 May 1993, p. B7.

201 UNESCO *UN Demographic Survey* (1982–86 average figures), cited in *The Economist World Atlas and Almanac*, 1989, p. 114.

203 *top, The New York Times*, 20 October 1991, sec. 4, p. 3; *bottom*, Telerate Teletrac, cited in *The Wall Street Journal*, 2 November 1990, p. C1.

205 Ricklefs (1993, p. 441).

209, 211, 213 *top*, Natural Resources Defense Council.

213 *bottom*, Morgan Stanley Capital International, cited in *The Economist World Atlas and Almanac*, 1989, p. 90.

215 *top*, Weisberg (1980), cited in Chambers et al. (1983, p. 371); *bottom*, EIU *Country Report*, cited in *The Economist World Atlas and Almanac*, 1989, p. 267.

217 *top, UN Energy Statistics Yearbook, Demographic Yearbook* (1985 figures), cited in *The Economist World Atlas and Almanac*, 1989, p. 95; *bottom*, UNESCO *UN Demographic Survey* (1982–86 average figures), cited in *The Economist World Atlas and Almanac*, 1989, p. 114.

219 *top*, EIU *Country Profile*, cited in *The Economist World Atlas and Almanac*, 1989, p. 162; *bottom*, Morgan Stanley Capital International, cited in *The Economist World Atlas and Almanac*, 1989, p. 90.

221 *bottom*, Morgan Stanley Capital International, cited in *The Economist World Atlas and Almanac*, 1989, p. 90.

223 *top and middle*, Food and Agricultural Organizations 8.14 (1979–81 average), cited in *The Economist World Atlas and Almanac*, 1989, p. 106.

225 Weisberg (1980), cited in Chambers et al. (1983, p. 371).

227   *top*, Weisberg (1980), cited in Chambers et al. (1983, 8.19 p. 371); *bottom*, *Two nations: Black and white, separate, hostile, unequal* (p. 98) by A. Hacker, 1992, New York: Scribner's, cited in *Newsweek*, 23 March 1992, p. 61.

229   *top*, U.S. Department of Transportation, cited in *The Economist*, 10 March 1990, p. 73; *bottom*, U.S. Department of Commerce, cited in *Newsweek*, 2 March 1992, p. 38.

231   *bottom*, Kosslyn, S. M., Ball, T. M., & Reiser, B. J. (1978). Visual images preserve metric spatial information: *Evidence from studies of image scanning. Journal of Experimental Psychology: Human Perception and Performance, 4,* 47–60.

233   *top*, Kosslyn (1976).

235   U.S. Department of Commerce, cited in *Newsweek*, 23 March 1992, p. 48.

index